CAMBRIDGE LIBRA

Books of enduring sc

Philosophy

This series contains both philosophical texts and critical essays about philosophy, concentrating especially on works originally published in the eighteenth and nineteenth centuries. It covers a broad range of topics including ethics, logic, metaphysics, aesthetics, utilitarianism, positivism, scientific method and political thought. It also includes biographies and accounts of the history of philosophy, as well as collections of papers by leading figures. In addition to this series, primary texts by ancient philosophers, and works with particular relevance to philosophy of science, politics or theology, may be found elsewhere in the Cambridge Library Collection.

Collected Essays

Known as 'Darwin's Bulldog', the biologist Thomas Henry Huxley (1825–95) was a tireless supporter of the evolutionary theories of his friend Charles Darwin. Huxley also made his own significant scientific contributions, and he was influential in the development of science education despite having had only two years of formal schooling. He established his scientific reputation through experiments on aquatic life carried out during a voyage to Australia while working as an assistant surgeon in the Royal Navy; ultimately he became President of the Royal Society (1883–5). Throughout his life Huxley struggled with issues of faith, and he coined the term 'agnostic' to describe his beliefs. This nine-volume collection of Huxley's essays, which he edited and published in 1893–4, demonstrates the wide range of his intellectual interests. Volume 9 includes lectures delivered at Oxford University on the relationship between scientific thought and ethical problems.

Cambridge University Press has long been a pioneer in the reissuing of out-of-print titles from its own backlist, producing digital reprints of books that are still sought after by scholars and students but could not be reprinted economically using traditional technology. The Cambridge Library Collection extends this activity to a wider range of books which are still of importance to researchers and professionals, either for the source material they contain, or as landmarks in the history of their academic discipline.

Drawing from the world-renowned collections in the Cambridge University Library, and guided by the advice of experts in each subject area, Cambridge University Press is using state-of-the-art scanning machines in its own Printing House to capture the content of each book selected for inclusion. The files are processed to give a consistently clear, crisp image, and the books finished to the high quality standard for which the Press is recognised around the world. The latest print-on-demand technology ensures that the books will remain available indefinitely, and that orders for single or multiple copies can quickly be supplied.

The Cambridge Library Collection will bring back to life books of enduring scholarly value (including out-of-copyright works originally issued by other publishers) across a wide range of disciplines in the humanities and social sciences and in science and technology.

Collected Essays

VOLUME 9: EVOLUTION AND ETHICS

THOMAS HENRY HUXLEY

CAMBRIDGE
UNIVERSITY PRESS

CAMBRIDGE UNIVERSITY PRESS

Cambridge, New York, Melbourne, Madrid, Cape Town,
Singapore, São Paolo, Delhi, Tokyo, Mexico City

Published in the United States of America by Cambridge University Press, New York

www.cambridge.org
Information on this title: www.cambridge.org/9781108040594

© in this compilation Cambridge University Press 2011

This edition first published 1894
This digitally printed version 2011

ISBN 978-1-108-04059-4 Paperback

COLLECTED ESSAYS

By T. H. HUXLEY

VOLUME IX

EVOLUTION & ETHICS

AND OTHER ESSAYS

BY

THOMAS H. HUXLEY

London
MACMILLAN AND CO.
1894

RICHARD CLAY AND SONS, LIMITED,
LONDON AND BUNGAY.

PREFACE

THE discourse on "Evolution and Ethics," reprinted in the first half of the present volume, was
delivered before the University of Oxford, as the
second of the annual lectures founded by Mr.
Romanes: whose name I may not write without
deploring the untimely death, in the flower of his
age, of a friend endeared to me, as to so many
others, by his kindly nature; and justly valued by
all his colleagues for his powers of investigation
and his zeal for the advancement of knowledge.
I well remember, when Mr. Romanes' early work
came into my hands, as one of the secretaries
of the Royal Society, how much I rejoiced in
the accession to the ranks of the little army of
workers in science of a recruit so well qualified
to take a high place among us.

It was at my friend's urgent request that I
agreed to undertake the lecture, should I be
honoured with an official proposal to give it, though
I confess not without misgivings, if only on

account of the serious fatigue and hoarseness
which public speaking has for some years caused
me; while I knew that it would be my fate to
follow the most accomplished and facile orator of
our time, whose indomitable youth is in no matter
more manifest than in his penetrating and musical voice. A certain saying about comparisons
intruded itself somewhat importunately.

And even if I disregarded the weakness of my
body in the matter of voice, and that of my mind
in the matter of vanity, there remained a third
difficulty. For several reasons, my attention,
during a number of years, has been much directed
to the bearing of modern scientific thought on the
problems of morals and of politics, and I did not
care to be diverted from that topic. Moreover, I
thought it the most important and the worthiest
which, at the present time, could engage the attention even of an ancient and renowned University.

But it is a condition of the Romanes foundation
that the lecturer shall abstain from treating of
either Religion or Politics; and it appeared to me
that, more than most, perhaps, I was bound to
act, not merely up to the letter, but in the spirit,
of that prohibition. Yet Ethical Science is, on
all sides, so entangled with Religion and Politics,
that the lecturer who essays to touch the former
without coming into contact with either of the
latter, needs all the dexterity of an egg-dancer;
and may even discover that his sense of clearness

and his sense of propriety come into conflict, by
no means to the advantage of the former.

I had little notion of the real magnitude of
these difficulties when I set about my task; but I
am consoled for my pains and anxiety by observing
that none of the multitudinous criticisms with
which I have been favoured and, often, instructed,
find fault with me on the score of having strayed
out of bounds.

Among my critics there are not a few to whom
I feel deeply indebted for the careful attention
which they have given to the exposition thus
hampered; and further weakened, I am afraid, by
my forgetfulness of a maxim touching lectures of
a popular character, which has descended to me
from that prince of lecturers, Mr. Faraday. He
was once asked by a beginner, called upon to
address a highly select and cultivated audience,
what he might suppose his hearers to know
already. Whereupon the past master of the art of
exposition emphatically replied " Nothing ! "

To my shame as a retired veteran, who has all
his life profited by this great precept of lec-
turing strategy, I forgot all about it just when
it would have been most useful. I was fatuous
enough to imagine that a number of propositions,
which I thought established, and which, in fact, I
had advanced without challenge on former occasions, needed no repetition.

I have endeavoured to repair my error by

prefacing the lecture with some matter—chiefly
elementary or recapitulatory—to which I have
given the title of "Prolegomena." I wish I could
have hit upon a heading of less pedantic aspect
which would have served my purpose; and if it
be urged that the new building looks over large
for the edifice to which it is added, I can only
plead the precedent of the ancient architects,
who always made the adytum the smallest part
of the temple.

If I had attempted to reply in full to the
criticisms to which I have referred, I know not
what extent of ground would have been covered
by my *pronaos*. All I have endeavoured to do,
at present, is to remove that which seems to
have proved a stumbling-block to many—namely,
the apparent paradox that ethical nature, while
born of cosmic nature, is necessarily at enmity
with its parent. Unless the arguments set forth
in the Prolegomena, in the simplest language at
my command, have some flaw which I am unable
to discern, this seeming paradox is a truth, as
great as it is plain, the recognition of which
is fundamental for the ethical philosopher.

We cannot do without our inheritance from the
forefathers who were the puppets of the cosmic
process; the society which renounces it must
be destroyed from without. Still less can we do
with too much of it; the society in which it
dominates must be destroyed from within.

The motive of the drama of human life is
the necessity, laid upon every man who comes
into the world, of discovering the mean between
self-assertion and self-restraint suited to his
character and his circumstances. And the eternally tragic aspect of the drama lies in this:
that the problem set before us is one the elements of which can be but imperfectly known,
and of which even an approximately right solution
rarely presents itself, until that stern critic, aged
experience, has been furnished with ample justification for venting his sarcastic humour upon the
irreparable blunders we have already made.

I have reprinted the letters on the "Darkest
England" scheme, published in the "Times" of
December 1890 and January 1891; and subsequently issued, with additions, as a pamphlet, under
the title of "Social Diseases and Worse Remedies;"
because, although the clever attempt to rush
the country on behalf of that scheme has been
balked, Mr. Booth's standing army remains afoot,
retaining all the capacities for mischief which are
inherent in its constitution. I am desirous that
this fact should be kept steadily in view; and
that the moderation of the clamour of the drums
and trumpets should not lead us to forget the
existence of a force, which, in bad hands, may, at
any time, be used for bad purposes.

In 1892, a Committee was "formed for the pur-

pose of investigating the manner in which the
moneys, subscribed in response to the appeal made
in the book entitled ' In Darkest England and the
Way out,' have been expended." The members of
this body were gentlemen in whose competency
and equity every one must have complete confidence; and in December 1892 they published
a report in which they declare that, " with the
exception of the sums expended on the ' barracks '
at Hadleigh," the moneys in question have been
" devoted only to the objects and expended in the
methods set out in that appeal, and to and in no
others."

Nevertheless, their final conclusion runs as
follows: " (4) That whilst the invested property,
real and personal, resulting from such Appeal is
so vested and controlled by the Trust of the Deed
of January 30th, 1891, that any application of it
to purposes other than those declared in the deed
by any ' General' of the Salvation Army would
amount to a breach of trust, and would subject him
to the proceedings of a civil and criminal character,
before mentioned in the Report, *adequate legal
safeguards do not at present exist to prevent the
misapplication of such property.*"

The passage I have italicised forms part of a
document dated December 19th, 1892. It follows,
that, even after the Deed of January 30th, 1891,
was executed, " adequate legal safeguards " " to
prevent the misapplication of the property " did

not exist. What then was the state of things, up to a week earlier, that is on January 22nd, 1891, when my twelfth and last letter appeared in the "Times"? A better justification for what I have said about the want of adequate security for the proper administration of the funds intrusted to Mr. Booth could not be desired, unless it be that which is to be found in the following passages of the Report (pp. 36 and 37):—

"It is possible that a 'General' may be forgetful of his duty, and sell property and appropriate the proceeds to his own use, or to meeting the general liabilities of the Salvation Army. As matters now stand, he, and he alone, would have control over such a sale. Against such possibilities it appears to the Committee to be reasonable that some check should be imposed."

Once more let it be remembered that this opinion, given under the hand of Sir Henry James, was expressed by the Committee, with the Trust Deed of 1891, which has been so sedulously flaunted before the public, in full view.

The Committee made a suggestion for the improvement of this very unsatisfactory state of things; but the exact value set upon it by the suggestors should be carefully considered (p. 37).

"The Committee are fully aware that if the views thus expressed are carried out, the safeguards and checks created will not be sufficient for all purposes absolutely to prevent possible dealing with the property and moneys, inconsistent with the purposes to which they are intended to be devoted."

In fact, they are content to express the very modest hope that "if the suggestion made be acted upon, some hindrance will thereby be placed in the way of any one acting dishonestly in respect of the disposal of the property and moneys referred to."

I do not know, and, under the circumstances, I cannot say I much care, whether the suggestions of the Committee have, or have not, been acted upon. Whether or not, the fact remains, that an unscrupulous "General" will have a pretty free hand, notwithstanding "some" hindrance.

Thus, the judgment of the highly authoritative, and certainly not hostile, Committee of 1892, upon the issues with which they concerned themselves is hardly such as to inspire enthusiastic confidence. And it is further to be borne in mind that they carefully excluded from their duties "any examination of the principles, government, teaching, or methods of the Salvation Army as a religious organisation, or of its affairs" except so far as they related to the administration of the moneys collected by the "Darkest England" appeal.

Consequently, the most important questions discussed in my letters were not in any way touched by the Committee. Even if their report had been far more favourable to the "Darkest England" scheme than it is; if it had really assured the contributors that the funds raised were fully secured against malversation; the objections, on social and political grounds, to Mr. Booth's despotic organization, with its thousands of docile satellites pledged to blind obedience, set forth in the letters, would be in no degree weakened. The "sixpennyworth of good" would still be outweighed by the "shilling'sworth of harm"; if indeed the relative worth, or unworth, of the latter should not be rated in pounds rather than in shillings.

What would one not give for the opinion of the financial members of the Committee about the famous Bank; and that of the legal experts about the proposed "tribunes of the people"?

HODESLEA, EASTBOURNE,
July 1894.

CONTENTS

I

EVOLUTION AND ETHICS

PROLEGOMENA

[1894]

I

IT may be safely assumed that, two thousand
years ago, before Cæsar set foot in southern
Britain, the whole country-side visible from the
windows of the room in which I write, was in
what is called "the state of nature." Except, it may
be, by raising a few sepulchral mounds, such as
those which still, here and there, break the flowing
contours of the downs, man's hands had made no
mark upon it; and the thin veil of vegetation
which overspread the broad-backed heights and
the shelving sides of the coombs was unaffected
by his industry. The native grasses and weeds,
the scattered patches of gorse, contended with one
another for the possession of the scanty surface
soil; they fought against the droughts of summer,

the frosts of winter, and the furious gales which swept, with unbroken force, now from the Atlantic, and now from the North Sea, at all times of the year ; they filled up, as they best might, the gaps made in their ranks by all sorts of underground and overground animal ravagers. One year with another, an average population, the floating balance of the unceasing struggle for existence among the indigenous plants, maintained itself. It is as little to be doubted, that an essentially similar state of nature prevailed, in this region, for many thousand years before the coming of Cæsar ; and there is no assignable reason for denying that it might continue to exist through an equally prolonged futurity, except for the intervention of man.

Reckoned by our customary standards of duration, the native vegetation, like the " everlasting hills " which it clothes, seems a type of permanence. The little Amarella Gentians, which abound in some places to-day, are the descendants of those that were trodden underfoot by the prehistoric savages who have left their flint tools about, here and there ; and they followed ancestors which, in the climate of the glacial epoch, probably flourished better than they do now. Compared with the long past of this humble plant, all the history of civilized men is but an episode.

Yet nothing is more certain than that, measured by the liberal scale of time-keeping of the universe, this present state of nature, however it may seem

to have gone and to go on for ever, is but a
fleeting phase of her infinite variety; merely the
last of the series of changes which the earth's surface has undergone in the course of the millions of
years of its existence. Turn back a square foot of
the thin turf, and the solid foundation of the land,
exposed in cliffs of chalk five hundred feet high on
the adjacent shore, yields full assurance of a time
when the sea covered the site of the "everlasting
hills"; and when the vegetation of what land lay
nearest, was as different from the present Flora of
the Sussex downs, as that of Central Africa now is.[1]
No less certain is it that, between the time during
which the chalk was formed and that at which the
original turf came into existence, thousands of
centuries elapsed, in the course of which, the state
of nature of the ages during which the chalk was
deposited, passed into that which now is, by
changes so slow that, in the coming and going of
the generations of men, had such witnessed them,
the contemporary conditions would have seemed
to be unchanging and unchangeable.

But it is also certain that, before the deposition
of the chalk, a vastly longer period had elapsed,
throughout which it is easy to follow the traces
of the same process of ceaseless modification and
of the internecine struggle for existence of living
things; and that even when we can get no further

[1] See "On a piece of Chalk" in the preceding volume of these
Essays (vol. viii. p. 1).

back, it is not because there is any reason to think
we have reached the beginning, but because the
trail of the most ancient life remains hidden, or
has become obliterated.

Thus that state of nature of the world of plants,
which we began by considering, is far from possess-
ing the attribute of permanence. Rather its very
essence is impermanence. It may have lasted
twenty or thirty thousand years, it may last for
twenty or thirty thousand years more, without
obvious change; but, as surely as it has followed
upon a very different state, so it will be followed
by an equally different condition. That which
endures is not one or another association of living
forms, but the process of which the cosmos is the
product, and of which these are among the transitory expressions. And in the living world, one of
the most characteristic features of this cosmic process is the struggle for existence, the competition
of each with all, the result of which is the selection,
that is to say, the survival of those forms which,
on the whole, are best adapted to the conditions
which at any period obtain; and which are, therefore, in that respect, and only in that respect, the
fittest.[1] The acme reached by the cosmic process

---

[1] That every theory of evolution must be consistent not
merely with progressive development, but with indefinite
persistence in the same condition and with retrogressive modification, is a point which I have insisted upon repeatedly from
the year 1862 till now. See *Collected Essays*, vol. ii. pp. 461–89 ;
vol. iii. p. 33 ; vol. viii. p. 304. In the address on "Geological

in the vegetation of the downs is seen in the turf, with its weeds and gorse. Under the conditions, they have come out of the struggle victorious; and, by surviving, have proved that they are the fittest to survive. That the state of nature, at any time, is a temporary phase of a process of incessant change, which has been going on for innumerable ages, appears to me to be a proposition as well established as any in modern history. Paleontology assures us, in addition, that the ancient philosophers who, with less reason, held the same doctrine, erred in supposing that the phases formed a cycle, exactly repeating the past, exactly foreshadowing the future, in their rotations. On the contrary, it furnishes us with conclusive reasons for thinking that, if every link in the ancestry of these humble indigenous plants had been preserved and were accessible to us, the whole would present a converging series of forms of gradually diminishing complexity, until, at some period in the history of the earth, far more remote than any of which organic remains have yet been discovered, they would merge in those low groups among which the boundaries between animal and vegetable life become effaced.[1]

Contemporaneity and Persistent Types " (1862), the paleontological proofs of this proposition were, I believe, first set forth.

[1] "On the Border Territory between the Animal and the Vegetable Kingdoms," Essays, vol. viii. p. 162.

The word " evolution," now generally applied to
the cosmic process, has had a singular history, and
is used in various senses.[1] Taken in its popular
signification it means progressive development,
that is, gradual change from a condition of
relative uniformity to one of relative complexity;
but its connotation has been widened to include
the phenomena of retrogressive metamorphosis,
that is, of progress from a condition of relative
complexity to one of relative uniformity.

As a natural process, of the same character as
the development of a tree from its seed, or of a
fowl from its egg, evolution excludes creation and
all other kinds of supernatural intervention. As
the expression of a fixed order, every stage of
which is the effect of causes operating according
to definite rules, the conception of evolution no
less excludes that of chance. It is very desirable
to remember that evolution is not an explanation
of the cosmic process, but merely a generalized
statement of the method and results of that process. And, further, that, if there is proof that
the cosmic process was set going by any agent,
then that agent will be the creator of it and of all
its products, although supernatural intervention
may remain strictly excluded from its further
course.

So far as that limited revelation of the nature of
things, which we call scientific knowledge, has

[1] See "Evolution in Biology," Essays, vol. ii. p. 187.

yet gone, it tends, with constantly increasing emphasis, to the belief that, not merely the world of plants, but that of animals; not merely living things, but the whole fabric of the earth ; not merely our planet, but the whole solar system ; not merely our star and its satellites, but the millions of similar bodies which bear witness to the order which pervades boundless space, and has endured through boundless time ; are all working out their predestined courses of evolution.

With none of these have I anything to do, at present, except with that exhibited by the forms of life which tenant the earth. All plants and animals exhibit the tendency to vary, the causes of which have yet to be ascertained ; it is the tendency of the conditions of life, at any given time, while favouring the existence of the variations best adapted to them, to oppose that of the rest and thus to exercise selection ; and all living things tend to multiply without limit, while the means of support are limited ; the obvious cause of which is the production of offspring more numerous than their progenitors, but with equal expectation of life in the actuarial sense. Without the first tendency there could be no evolution. Without the second, there would be no good reason why one variation should disappear and another take its place ; that is to say, there would be no selection. Without the third, the struggle for existence, the agent of the selective process in the state of nature, would vanish.[1]

Granting the existence of these tendencies, all the known facts of the history of plants and of animals may be brought into rational correlation. And this is more than can be said for any other hypothesis that I know of. Such hypotheses, for example, as that of the existence of a primitive, orderless chaos; of a passive and sluggish eternal matter moulded, with but partial success, by archetypal ideas; of a brand-new world-stuff suddenly created and swiftly shaped by a supernatural power; receive no encouragement, but the contrary, from our present knowledge. That our earth may once have formed part of a nebulous cosmic magma is certainly possible, indeed seems highly probable; but there is no reason to doubt that order reigned there, as completely as amidst what we regard as the most finished works of nature or of man.[2] The faith which is born of knowledge, finds its object in an eternal order, bringing forth ceaseless change, through endless time, in endless space; the manifestations of the cosmic energy alternating between phases of potentiality and phases of explication. It may be that, as Kant suggests,[3] every cosmic

---

[1] *Collected Essays*, vol. ii. *passim.*
[2] *Ibid.*, vol. iv. p. 138 ; vol. v. pp. 71-73.
[3] *Ibid.*, vol. viii. p. 321.

magma predestined to evolve into a new world,
has been the no less predestined end of a vanished predecessor.

II

Three or four years have elapsed since the state
of nature, to which I have referred, was brought
to an end, so far as a small patch of the soil is
concerned, by the intervention of man. The
patch was cut off from the rest by a wall; within
the area thus protected, the native vegetation was,
as far as possible, extirpated; while a colony of
strange plants was imported and set down in its
place. In short, it was made into a garden. At
the present time, this artificially treated area
presents an aspect extraordinarily different from
that of so much of the land as remains in the
state of nature, outside the wall. Trees, shrubs,
and herbs, many of them appertaining to the
state of nature of remote parts of the globe,
abound and flourish. Moreover, considerable
quantities of vegetables, fruits, and flowers are
produced, of kinds which neither now exist, nor
have ever existed, except under conditions such as
obtain in the garden; and which, therefore, are as
much works of the art of man as the frames and
glass-houses in which some of them are raised.
That the "state of Art," thus created in the
state of nature by man, is sustained by
and dependent on him, would at once become

apparent, if the watchful supervision of the gardener were withdrawn, and the antagonistic influences of the general cosmic process were no longer
sedulously warded off, or counteracted. The walls
and gates would decay; quadrupedal and bipedal
intruders would devour and tread down the useful
and beautiful plants; birds, insects, blight, and
mildew would work their will; the seeds of the
native plants, carried by winds or other agencies,
would immigrate, and in virtue of their long-
earned special adaptation to the local conditions,
these despised native weeds would soon choke
their choice exotic rivals. A century or two
hence, little beyond the foundations of the wall
and of the houses and frames would be left, in
evidence of the victory of the cosmic powers at
work in the state of nature, over the temporary
obstacles to their supremacy, set up by the art of
the horticulturist.

It will be admitted that the garden is as much
a work of art,[1] or artifice, as anything that can be
mentioned. The energy localised in certain human
bodies, directed by similarly localised intellects,
has produced a collocation of other material bodies
which could not be brought about in the state of
nature. The same proposition is true of all the

---

[1] The sense of the term "Art" is becoming narrowed;
"work of Art" to most people means a picture, a statue, or a
piece of *bijouterie ;* by way of compensation "artist" has included in its wide embrace cooks and ballet girls, no less than
painters and sculptors.

works of man's hands, from a flint implement to
a cathedral or a chronometer; and it is because
it is true, that we call these things artificial,
term them works of art, or artifice, by way of
distinguishing them from the products of the cosmic
process, working outside man, which we call
natural, or works of nature. The distinction thus
drawn between the works of nature and those of
man, is universally recognised; and it is, as I
conceive, both useful and justifiable.

III

No doubt, it may be properly urged that
the operation of human energy and intelligence,
which has brought into existence and maintains
the garden, by what I have called "the horticultural process," is, strictly speaking, part and parcel
of the cosmic process. And no one could more
readily agree to that proposition than I. In fact,
I do not know that any one has taken more
pains than I have, during the last thirty years, to
insist upon the doctrine, so much reviled in the
early part of that period, that man, physical,
intellectual, and moral, is as much a part of
nature, as purely a product of the cosmic process,
as the humblest weed.[1]

But if, following up this admission, it is urged

[1] See "Man's Place in Nature," *Collected Essays*, vol. vii., and
"On the Struggle for Existence in Human Society" (1888), below.

that, such being the case, the cosmic process cannot be in antagonism with that horticultural process which is part of itself—I can only reply, that
if the conclusion that the two are antagonistic
is logically absurd, I am sorry for logic, because,
as we have seen, the fact is so. The garden is in
the same position as every other work of man's
art; it is a result of the cosmic process working
through and by human energy and intelligence;
and, as is the case with every other artificial
thing set up in the state of nature, the influences of the latter are constantly tending to break
it down and destroy it. No doubt, the Forth bridge
and an ironclad in the offing, are, in ultimate resort, products of the cosmic process; as much so as
the river which flows under the one, or the sea-
water on which the other floats. Nevertheless,
every breeze strains the bridge a little, every tide
does something to weaken its foundations; every
change of temperature alters the adjustment of
its parts, produces friction and consequent wear
and tear. From time to time, the bridge must be
repaired, just as the ironclad must go into dock;
simply because nature is always tending to re-
claim that which her child, man, has borrowed
from her and has arranged in combinations which
are not those favoured by the general cosmic
process.

Thus, it is not only true that the cosmic
energy, working through man upon a portion of the

plant world, opposes the same energy as it works through the state of nature, but a similar antagonism is everywhere manifest between the artificial and the natural. Even in the state of nature itself, what is the struggle for existence but the antagonism of the results of the cosmic process in the region of life, one to another? [1]

IV

Not only is the state of nature hostile to the state of art of the garden; but the principle of the horticultural process, by which the latter is created and maintained, is antithetic to that of the cosmic process. The characteristic feature of the latter is the intense and unceasing competition of the struggle for existence. The characteristic of the former is the elimination of that struggle, by the removal of the conditions which give rise to it. The tendency of the cosmic process is to bring about the adjustment of the forms of plant life to the current conditions; the tendency of the horticultural process is the adjustment of the conditions to the needs of the forms of plant life which the gardener desires to raise.

The cosmic process uses unrestricted multiplica-

---

[1] Or to put the case still more simply. When a man lays hold of the two ends of a piece of string and pulls them, with intent to break it, the right arm is certainly exerted in antagonism to the left arm; yet both arms derive their energy from the same original source.

tion as the means whereby hundreds compete for
the place and nourishment adequate for one; it
employs frost and drought to cut off the weak
and unfortunate; to survive, there is need not
only of strength, but of flexibility and of good
fortune.

The gardener, on the other hand, restricts
multiplication; provides that each plant shall
have sufficient space and nourishment; protects
from frost and drought; and, in every other way,
attempts to modify the conditions, in such a
manner as to bring about the survival of those
forms which most nearly approach the standard
of the useful, or the beautiful, which he has in
his mind.

If the fruits and the tubers, the foliage and
the flowers thus obtained, reach, or sufficiently
approach, that ideal, there is no reason why the
status quo attained should not be indefinitely prolonged. So long as the state of nature remains
approximately the same, so long will the energy
and intelligence which created the garden suffice to
maintain it. However, the limits within which this
mastery of man over nature can be maintained are
narrow. If the conditions of the cretaceous epoch
returned, I fear the most skilful of gardeners would
have to give up the cultivation of apples and
gooseberries; while, if those of the glacial period
once again obtained, open asparagus beds would
be superfluous, and the training of fruit trees

against the most favourable of south walls, a waste of time and trouble.

But it is extremely important to note that, the state of nature remaining the same, if the produce does not satisfy the gardener, it may be made to approach his ideal more closely. Although the struggle for existence may be at end, the possibility of progress remains. In discussions on these topics, it is often strangely forgotten that the essential conditions of the modification, or evolution, of living things are variation and hereditary transmission. Selection is the means by which certain variations are favoured and their progeny preserved. But the struggle for existence is only one of the means by which selection may be effected. The endless varieties of cultivated flowers, fruits, roots, tubers, and bulbs are not products of selection by means of the struggle for existence, but of direct selection, in view of an ideal of utility or beauty. Amidst a multitude of plants, occupying the same station and subjected to the same conditions, in the garden, varieties arise. The varieties tending in a given direction are preserved, and the rest are destroyed. And the same process takes place among the varieties until, for example, the wild kale becomes a cabbage, or the wild *Viola tricolor* a prize pansy.

V

The process of colonization presents analogies
to the formation of a garden which are highly
instructive. Suppose a shipload of English
colonists sent to form a settlement, in such a
country as Tasmania was in the middle of the last
century. On landing, they find themselves in the
midst of a state of nature, widely different from
that left behind them in everything but the most
general physical conditions. The common plants,
the common birds and quadrupeds, are as totally
distinct as the men from anything to be seen on
the side of the globe from which they come.
The colonists proceed to put an end to this state
of things over as large an area as they desire to
occupy. They clear away the native vegetation,
extirpate or drive out the animal population, so
far as may be necessary, and take measures to
defend themselves from the re-immigration of
either. In their place, they introduce English
grain and fruit trees; English dogs, sheep, cattle,
horses; and English men; in fact, they set up a
new Flora and Fauna and a new variety of mankind,
within the old state of nature. Their farms and
pastures represent a garden on a great scale, and
themselves the gardeners who have to keep it up,
in watchful antagonism to the old *régime.* Considered as a whole, the colony is a composite unit
introduced into the old state of nature; and,

thenceforward, a competitor in the struggle for existence, to conquer or be vanquished.

Under the conditions supposed, there is no doubt of the result, if the work of the colonists be carried out energetically and with intelligent combination of all their forces. On the other hand, if they are slothful, stupid, and careless; or if they waste their energies in contests with one another, the chances are that the old state of nature will have the best of it. The native savage will destroy the immigrant civilized man; of the English animals and plants some will be extirpated by their indigenous rivals, others will pass into the feral state and themselves become components of the state of nature. In a few decades, all other traces of the settlement will have vanished.

VI

Let us now imagine that some administrative authority, as far superior in power and intelligence to men, as men are to their cattle, is set over the colony, charged to deal with its human elements in such a manner as to assure the victory of the settlement over the antagonistic influences of the state of nature in which it is set down. He would proceed in the same fashion as that in which the gardener dealt with his garden. In the first place, he would, as far as possible, put a

stop to the influence of external competition by
thoroughly extirpating and excluding the native
rivals, whether men, beasts, or plants. And
our administrator would select his human agents,
with a view to his ideal of a successful colony,
just as the gardener selects his plants with a view
to his ideal of useful or beautiful products.

In the second place, in order that no struggle
for the means of existence between these human
agents should weaken the efficiency of the corporate whole in the battle with the state of
nature, he would make arrangements by which
each would be provided with those means; and
would be relieved from the fear of being deprived
of them by his stronger or more cunning fellows.
Laws, sanctioned by the combined force of the
colony, would restrain the self-assertion of each
man within the limits required for the maintenance of peace. In other words, the cosmic struggle
for existence, as between man and man, would be
rigorously suppressed ; and selection, by its means,
would be as completely excluded as it is from
the garden.

At the same time, the obstacles to the full
development of the capacities of the colonists
by other conditions of the state of nature
than those already mentioned, would be re-
moved by the creation of artificial conditions of
existence of a more favourable character. Protection against extremes of heat and cold would

be afforded by houses and clothing; drainage and irrigation works would antagonise the effects of excessive rain and excessive drought; roads, bridges, canals, carriages, and ships would overcome the natural obstacles to locomotion and transport; mechanical engines would supplement the natural strength of men and of their draught animals; hygienic precautions would check, or remove, the natural causes of disease. With every step of this progress in civilization, the colonists would become more and more independent of the state of nature; more and more, their lives would be conditioned by a state of art. In order to attain his ends, the administrator would have to avail himself of the courage, industry, and co-operative intelligence of the settlers; and it is plain that the interest of the community would be best served by increasing the proportion of persons who possess such qualities, and diminishing that of persons devoid of them. In other words, by selection directed towards an ideal.

Thus the administrator might look to the establishment of an earthly paradise, a true garden of Eden, in which all things should work together towards the well-being of the gardeners: within which the cosmic process, the coarse struggle for existence of the state of nature, should be abolished; in which that state should be replaced by a state of art;

where every plant and every lower animal should be adapted to human wants, and would perish if human supervision and protection were withdrawn; where men themselves should have been selected, with a view to their efficiency as organs for the performance of the functions of a perfected society. And this ideal polity would have been brought about, not by gradually adjusting the men to the conditions around them, but by creating artificial conditions for them; not by allowing the free play of the struggle for existence, but by excluding that struggle; and by substituting selection directed towards the administrator's ideal for the selection it exercises.

VII

But the Eden would have its serpent, and a very subtle beast too. Man shares with the rest of the living world the mighty instinct of reproduction and its consequence, the tendency to multiply with great rapidity. The better the measures of the administrator achieved their object, the more completely the destructive agencies of the state of nature were defeated, the less would that multiplication be checked.

On the other hand, within the colony, the enforcement of peace, which deprives every man of the power to take away the means of existence from another, simply because he is the stronger, would have put an end to the struggle for existence between the colonists, and the competition for the commodities of existence, which would alone remain, is no check upon population.

Thus, as soon as the colonists began to multiply, the administrator would have to face the tendency to the reintroduction of the cosmic struggle into his artificial fabric, in consequence of the competition, not merely for the commodities, but for the means of existence. When the colony reached the limit of possible expansion, the surplus population must be disposed of somehow; or the fierce struggle for existence must recommence and destroy that peace, which is the fundamental condition of the maintenance of the state of art against the state of nature.

Supposing the administrator to be guided by purely scientific considerations, he would, like the gardener, meet this most serious difficulty by systematic extirpation, or exclusion, of the superfluous. The hopelessly diseased, the infirm aged, the weak or deformed in body or in mind, the excess of infants born, would be put away, as the gardener pulls up defective and superfluous plants, or the breeder destroys undesirable cattle. Only the strong and the healthy, carefully matched, with a view to the progeny best adapted to the purposes of the administrator, would be permitted to perpetuate their kind.

VIII

Of the more thoroughgoing of the multitudinous
attempts to apply the principles of cosmic evolution, or what are supposed to be such, to social
and political problems, which have appeared of late
years, a considerable proportion appear to me to
be based upon the notion that human society is
competent to furnish, from its own resources, an
administrator of the kind I have imagined. The
pigeons, in short, are to be their own Sir John
Sebright.[1] A despotic government, whether individual or collective, is to be endowed with the
preternatural intelligence, and with what, I am
afraid, many will consider the preternatural ruth-
lessness, required for the purpose of carrying out
the principle of improvement by selection, with the
somewhat drastic thoroughness upon which the
success of the method depends. Experience certainly does not justify us in limiting the ruthless-
ness of individual "saviours of society"; and, on
the well-known grounds of the aphorism which
denies both body and soul to corporations, it seems
probable (indeed the belief is not without support
in history) that a collective despotism, a mob got
to believe in its own divine right by demagogic
missionaries, would be capable of more thorough

---

[1] Not that the conception of such a society is necessarily based
upon the idea of evolution. The Platonic state testifies to the
contrary.

work in this direction than any single tyrant, puffed up with the same illusion, has ever achieved. But intelligence is another affair. The fact that " saviours of society " take to that trade is evidence enough that they have none to spare. And such as they possess is generally sold to the capitalists of physical force on whose resources they depend. However, I doubt whether even the keenest judge of character, if he had before him a hundred boys and girls under fourteen, could pick out, with the least chance of success, those who should be kept, as certain to be serviceable members of the polity, and those who should be chloroformed, as equally sure to be stupid, idle, or vicious. The " points " of a good or of a bad citizen are really far harder to discern than those of a puppy or a short-horn calf ; many do not show themselves before the practical difficulties of life stimulate manhood to full exertion. And by that time the mischief is done. The evil stock, if it be one, has had time to multiply, and selection is nullified.

IX

I have other reasons for fearing that this logical ideal of evolutionary regimentation—this pigeon-fanciers' polity—is unattainable. In the absence of any such a severely scientific administrator as we have been dreaming of, human society is kept together by bonds of such a singular character, that the attempt to perfect society after his fashion would run serious risk of loosening them.

Social organization is not peculiar to men. Other societies, such as those constituted by bees and ants, have also arisen out of the advantage of co-operation in the struggle for existence; and their resemblances to, and their differences from, human society are alike instructive. The society formed by the hive bee fulfils the ideal of the communistic aphorism "to each according to his needs, from each according to his capacity." Within it, the struggle for existence is strictly limited. Queen, drones, and workers have each their allotted sufficiency of food; each performs the function assigned to it in the economy of the hive, and all contribute to the success of the whole co-operative society in its competition with rival collectors of nectar and pollen and with other enemies, in the state of nature without. In the same sense as the garden, or the colony, is a work of human art, the bee polity is a work of apiarian art, brought about by the cosmic process, working through the organization of the hymenopterous type.

Now this society is the direct product of an organic necessity, impelling every member of it to a course of action which tends to the good of the whole. Each bee has its duty and none

has any rights. Whether bees are susceptible
of feeling and capable of thought is a question
which cannot be dogmatically answered. As a
pious opinion, I am disposed to deny them more
than the merest rudiments of consciousness.[1] But
it is curious to reflect that a thoughtful drone
(workers and queens would have no leisure for
speculation) with a turn for ethical philosophy,
must needs profess himself an intuitive moralist
of the purest water. He would point out, with
perfect justice, that the devotion of the workers to
a life of ceaseless toil for a mere subsistence
wage, cannot be accounted for either by enlight-
ened selfishness, or by any other sort of utilitarian
motives; since these bees begin to work, without
experience or reflection, as they emerge from the
cell in which they are hatched. Plainly, an eternal and immutable principle, innate in each bee,
can alone account for the phenomena. On the
other hand, the biologist, who traces out all the
extant stages of gradation between solitary and
hive bees, as clearly sees in the latter, simply the
perfection of an automatic mechanism, hammered
out by the blows of the struggle for existence
upon the progeny of the former, during long ages
of constant variation.

[1] *Collected Essays*, vol. i., "Animal Automatism"; vol. v.,
"Prologue," pp. 45 *et seq.*

X

I see no reason to doubt that, at its origin, human society was as much a product of organic necessity as that of the bees.[1] The human family, to begin with, rested upon exactly the same conditions as those which gave rise to similar associations among animals lower in the scale. Further, it is easy to see that every increase in the duration of the family ties, with the resulting co-operation of a larger and larger number of descendants for protection and defence, would give the families in which such modification took place a distinct advantage over the others. And, as in the hive, the progressive limitation of the struggle for existence between the members of the family would involve increasing efficiency as regards outside competition.

But there is this vast and fundamental difference between bee society and human society. In the former, the members of the society are each organically predestined to the performance of one particular class of functions only. If they were endowed with desires, each could desire to perform none but those offices for which its organization specially fits it; and which, in view of the good of the whole, it is proper it should do. So long as a new queen does not make her appearance, rivalries and competition are absent from the bee polity.

[1] *Collected Essays*, vol. v., Prologue, pp. 50-54.

Among mankind, on the contrary, there is no such predestination to a sharply defined place in the social organism. However much men may differ in the quality of their intellects, the intensity of their passions, and the delicacy of their sensations, it cannot be said that one is fitted by his organization to be an agricultural labourer and nothing else, and another to be a landowner and nothing else. Moreover, with all their enormous differences in natural endowment, men agree in one thing, and that is their innate desire to enjoy the pleasures and to escape the pains of life; and, in short, to do nothing but that which it pleases them to do, without the least reference to the welfare of the society into which they are born. That is their inheritance (the reality at the bottom of the doctrine of original sin) from the long series of ancestors, human and semi-human and brutal, in whom the strength of this innate tendency to self-assertion was the condition of victory in the struggle for existence. That is the reason of the *aviditas vitæ* [1]—the insatiable hunger for enjoyment—of all mankind, which is one of the essential conditions of success in the war with the state of nature outside; and yet the sure agent of the destruction of society if allowed free play within.

The check upon this free play of self-assertion, or natural liberty, which is the necessary condition for the origin of human society, is the product

[1] See below. Romanes' Lecture, note 7.

of organic necessities of a different kind from
those upon which the constitution of the hive
depends. One of these is the mutual affection
of parent and offspring, intensified by the long
infancy of the human species. But the most
important   is   the   tendency,   so   strongly
developed in man, to reproduce in himself actions and feelings similar to, or correlated with,
those of other men. Man is the most con-
summate of all mimics in the animal world;
none but himself can draw or model; none comes
near him in the scope, variety, and exactness of
vocal imitation; none is such a master of gesture;
while he seems to be impelled thus to imitate
for the pure pleasure of it. And there is
no such another emotional chameleon. By a
purely reflex operation of the mind, we take
the hue of passion of those who are about us,
or, it may be, the complementary colour. It is
not by any conscious " putting one's self in the
place " of a joyful or a suffering person that the
state of mind we call sympathy usually arises;[1]
indeed, it is often contrary to one's sense of

---

[1] Adam Smith makes the pithy observation that the man
who sympathises with a woman in childbed, cannot be said
to put himself in her place. ("The Theory of the Moral Sentiments," Part vii. sec. iii. chap. i.) Perhaps there is more
humour than force in the example ; and, in spite of this
and other observations of the same tenor, I think that the
one defect of the remarkable work in which it occurs is that
it lays too much stress on conscious substitution, too little
on purely reflex sympathy.

right, and in spite of one's will, that "fellow-feeling makes us wondrous kind," or the reverse However complete may be the indifference to public opinion, in a cool, intellectual view, of the traditional sage, it has not yet been my fortune to meet with any actual sage who took its hostile manifestations with entire equanimity. Indeed, I doubt if the philosopher lives, or ever has lived, who could know himself to be heartily despised by a street boy without some irritation. And, though one cannot justify Haman for wishing to hang Mordecai on such a very high gibbet, yet, really, the consciousness of the Vizier of Ahasuerus, as he went in and out of the gate, that this obscure Jew had no respect for him, must have been very annoying.[1]

It is needful only to look around us, to see that the greatest restrainer of the anti-social tendencies of men is fear, not of the law, but of the opinion of their fellows. The conventions of honour bind men who break legal, moral, and religious bonds; and, while people endure the extremity of physical pain rather than part with life, shame drives the weakest to suicide.

Every forward step of social progress brings men

---

[1] Esther v. 9–13. ". . . but when Haman saw Mordecai in the king's gate, that he stood not up, nor moved for him, he was full of indignation against Mordecai. . . . And Haman told them of the glory of his riches. . . . and all the things wherein the king had promoted him. . . . Yet all this availeth me nothing, so long as I see Mordecai the Jew sitting at the king's gate." What a shrewd exposure of human weakness it is !

into closer relations with their fellows, and increases the importance of the pleasures and pains derived from sympathy. We judge the acts of others by our own sympathies, and we judge our own acts by the sympathies of others, every day and all day long, from childhood upwards, until associations, as indissoluble as those of language, are formed between certain acts and the feelings of approbation or disapprobation. It becomes impossible to imagine some acts without disapprobation, or others without approbation of the actor, whether he be one's self, or any one else. We come to think in the acquired dialect of morals. An artificial personality, the "man within," as Adam Smith [1] calls conscience, is built up beside the natural personality. He is the watchman of society, charged to restrain the anti-social tendencies of the natural man within the limits required by social welfare.

XI

I have termed this evolution of the feelings out of which the primitive bonds of human society are so largely forged, into the organized and personified sympathy we call conscience, the ethical process.[2] So far as it tends to

---

[1] "Theory of the Moral Sentiments," Part iii. chap. 3. *On the influence and authority of conscience.*

[2] Worked out, in its essential features, chiefly by Hartley and Adam Smith, long before the modern doctrine of evolution was thought of. See *Note* below, p. 45.

make any human society more efficient in
the struggle for existence with the state of
nature, or with other societies, it works in harmonious contrast with the cosmic process. But
it is none the less true that, since law and
morals are restraints upon the struggle for existence between men in society, the ethical
process is in opposition to the principle of the
cosmic process, and tends to the suppression of the
qualities best fitted for success in that struggle.[1]

It is further to be observed that, just as the self-
assertion, necessary to the maintenance of society
against the state of nature, will destroy that society
if it is allowed free operation within; so the self-
restraint, the essence of the ethical process, which
is no less an essential condition of the existence of
every polity, may, by excess, become ruinous to it.

Moralists of all ages and of all faiths, attending
only to the relations of men towards one another
in an ideal society, have agreed upon the
"golden rule," "Do as you would be done by."
In other words, let sympathy be your guide;
put yourself in the place of the man towards
whom your action is directed; and do to him
what you would like to have done to yourself
under the circumstances. However much one
may admire the generosity of such a rule of con-

---

[1] See the essay "On the Struggle for Existence in Human
Society" below; and *Collected Essays*, vol. i. p. 276, for Kant's
recognition of these facts.

duct; however confident one may be that average men may be thoroughly depended upon not to carry it out to its full logical consequences; it is nevertheless desirable to recognise the fact that these consequences are incompatible with the existence of a civil state, under any circumstances of this world which have obtained, or, so far as one can see, are, likely to come to pass.

For I imagine there can be no doubt that the great desire of every wrongdoer is to escape from the painful consequences of his actions. If I put myself in the place of the man who has robbed me, I find that I am possessed by an exceeding desire not to be fined or imprisoned; if in that of the man who has smitten me on one cheek, I contemplate with satisfaction the absence of any worse result than the turning of the other cheek for like treatment. Strictly observed, the "golden rule" involves the negation of law by the refusal to put it in motion against law-breakers; and, as regards the external relations of a polity, it is the refusal to continue the struggle for existence. It can be obeyed, even partially, only under the protection of a society which repudiates it. Without such shelter, the followers of the "golden rule" may indulge in hopes of heaven, but they must reckon with the certainty that other people will be masters of the earth.

What would become of the garden if the gar-

dener treated all the weeds and slugs and birds
and trespassers as he would like to be treated, if
he were in their place?

XII

Under the preceding heads, I have endeavoured
to represent in broad, but I hope faithful, outlines
the essential features of the state of nature and of
that cosmic process of which it is the outcome, so
far as was needful for my argument; I have con-
trasted with the state of nature the state of
art, produced by human intelligence and energy,
as it is exemplified by a garden; and I have shown
that the state of art, here and elsewhere, can be
maintained only by the constant counteraction
of the hostile influences of the state of nature.
Further, I have pointed out that the " horticultural
process" which thus sets itself against the " cosmic
process" is opposed to the latter in principle, in so
far as it tends to arrest the struggle for existence,
by restraining the multiplication which is one
of the chief causes of that struggle, and by
creating artificial conditions of life, better adapted
to the cultivated plants than are the conditions of
the state of nature.   And I have dwelt upon the
fact that, though the progressive modification,
which is the consequence of the struggle for
existence in the state of nature, is at an end,
such modification may still be effected by that

selection, in view of an ideal of usefulness, or of pleasantness, to man, of which the state of nature knows nothing.

I have proceeded to show that a colony, set down in a country in the state of nature, presents close analogies with a garden; and I have indicated the course of action which an administrator, able and willing to carry out horticultural principles, would adopt, in order to secure the success of such a newly formed polity, supposing it to be capable of indefinite expansion. In the contrary case, I have shown that difficulties must arise ; that the unlimited increase of the population over a limited area must, sooner or later, reintroduce into the colony that struggle for the means of existence between the colonists, which it was the primary object of the administrator to exclude, insomuch as it is fatal to the mutual peace which is the prime condition of the union of men in society.

I have briefly described the nature of the only radical cure, known to me, for the disease which would thus threaten the existence of the colony ; and, however regretfully, I have been obliged to admit that this rigorously scientific method of applying the principles of evolution to human society hardly comes within the region of practical politics ; not for want of will on the part of a great many people; but because, for one reason, there is no hope that mere human beings will ever possess enough intelligence to select the fittest. And I have adduced other grounds for arriving at the same conclusion.

I have pointed out that human society took its rise in the organic necessities expressed by imitation and by the sympathetic emotions; and that, in the struggle for existence with the state of nature and with other societies, as part of it, those in which men were thus led to close co-operation had a great advantage.[1] But, since each man retained more or less of the faculties common to all the rest, and especially a full share of the desire for unlimited self-gratification, the struggle for existence within society could only be gradually eliminated. So long as any of it remained, society continued to be an imperfect instrument of the struggle for existence and, consequently, was improvable by the selective influence of that struggle. Other things being alike, the tribe of savages in which order was best maintained; in which there was most security within the tribe and the most loyal mutual support outside it, would be the survivors.

I have termed this gradual strengthening of the social bond, which, though it arrests the struggle for existence inside society, up to a certain point improves the chances of society, as a corporate whole, in the cosmic struggle—the ethical process. I have endeavoured to show that, when the ethical process has advanced so

---

[1] *Collected Essays*, vol. v., Prologue, p. 52.

far as to secure every member of the society in
the possession of the means of existence, the
struggle for existence, as between man and man,
within that society is, *ipso facto*, at an end. And,
as it is undeniable that the most highly civilized
societies have substantially reached this position,
it follows that, so far as they are concerned, the
struggle for existence can play no important part
within them.[1] In other words, the kind of evolution which is brought about in the state of
nature cannot take place.

I have further shown cause for the belief that
direct selection, after the fashion of the horticulturist and the breeder, neither has played, nor
can play, any important part in the evolution
of society ; apart from other reasons, because I
do not see how such selection could be practised
without a serious weakening, it may be the destruction, of the bonds which hold society together.
It strikes me that men who are accustomed to
contemplate the active or passive extirpation of
the weak, the unfortunate, and the superfluous ;
who justify that conduct on the ground that it has
the sanction of the cosmic process, and is the only
way of ensuring the progress of the race ; who, if

---

[1] Whether the struggle for existence with the state of nature
and with other societies, so far as they stand in the relation of
the state of nature with it, exerts a selective influence upon modern
society, and in what direction, are questions not easy to
answer.   The problem of the effect of military and industrial
warfare upon those who wage it is very complicated.

they are consistent, must rank medicine among the black arts and count the physician a mischievous preserver of the unfit; on whose matrimonial undertakings the principles of the stud have the chief influence; whose whole lives, therefore, are an education in the noble art of suppressing natural affection and sympathy, are not likely to have any large stock of these commodities left. But, without them, there is no conscience, nor any restraint on the conduct of men, except the calculation of self-interest, the balancing of certain present gratifications against doubtful future pains; and experience tells us how much that is worth. Every day, we see firm believers in the hell of the theologians commit acts by which, as they believe when cool, they risk eternal punishment; while they hold back from those which are opposed to the sympathies of their associates.

XIII

That progressive modification of civilization which passes by the name of the "evolution of society," is, in fact, a process of an essentially different character, both from that which brings about the evolution of species, in the state of nature, and from that which gives rise to the evolution of varieties, in the state of art.

There can be no doubt that vast changes have taken place in English civilization since the reign

of the Tudors. But I am not aware of a particle
of evidence in favour of the conclusion that this
evolutionary process has been accompanied by any
modification of the physical, or the mental,
characters of the men who have been the subjects
of it. I have not met with any grounds for
suspecting that the average Englishmen of to-day
are sensibly different from those that Shakspere
knew and drew. We look into his magic mirror
of the Elizabethan age, and behold, nowise darkly,
the presentment of ourselves.

During these three centuries, from the reign of
Elizabeth to that of Victoria, the struggle for
existence between man and man has been so
largely restrained among the great mass of the
population (except for one or two short intervals
of civil war), that it can have had little, or no,
selective operation. As to anything comparable
to direct selection, it has been practised on so
small a scale that it may also be neglected.
The criminal law, in so far as by putting to death,
or by subjecting to long periods of imprisonment,
those who infringe its provisions, it prevents the
propagation of hereditary criminal tendencies;
and the poor-law, in so far as it separates married
couples, whose destitution arises from hereditary
defects of character, are doubtless selective agents
operating in favour of the non-criminal and the
more effective members of society. But the proportion of the population which they influence

is very small; and, generally, the hereditary criminal and the hereditary pauper have propagated their kind before the law affects them. In a large proportion of cases, crime and pauperism have nothing to do with heredity; but are the consequence, partly, of circumstances and, partly, of the possession of qualities, which, under different conditions of life, might have excited esteem and even admiration. It was a shrewd man of the world who, in discussing sewage problems, remarked that dirt is riches in the wrong place; and that sound aphorism has moral applications. The benevolence and open-handed generosity which adorn a rich man, may make a pauper of a poor one ; the energy and courage to which the successful soldier owes his rise, the cool and daring subtlety to which the great financier owes his fortune, may very easily, under unfavourable conditions, lead their possessors to the gallows, or to the hulks. Moreover, it is fairly probable that the children of a ' failure ' will receive from their other parent just that little modification of character which makes all the difference. I sometimes wonder whether people, who talk so freely about extirpating the unfit, ever dispassionately consider their own history. Surely, one must be very ' fit,' indeed, not to know of an occasion, or perhaps two, in one's life, when it would have been only too easy to qualify for a place among the ' unfit.'

In my belief the innate qualities, physical, intellectual, and moral, of our nation have remained substantially the same for the last four or five centuries. If the struggle for existence has affected us to any serious extent (and I doubt it) it has been, indirectly, through our military and industrial wars with other nations.

XIV

What is often called the struggle for existence in society (I plead guilty to having used the term too loosely myself), is a contest, not for the means of existence, but for the means of enjoyment. Those who occupy the first places in this practical competitive examination are the rich and the influential; those who fail, more or less, occupy the lower places, down to the squalid obscurity of the pauper and the criminal. Upon the most liberal estimate, I suppose the former group will not amount to two per cent. of the population. I doubt if the latter exceeds another two per cent.; but let it be supposed, for the sake of argument, that it is as great as five per cent.[1]

As it is only in the latter group that anything comparable to the struggle for existence in the state of nature can take place; as it is only

[1] Those who read the last Essay in this volume will not accuse me of wishing to attenuate the evil of the existence of this group, whether great or small.

among this twentieth of the whole people that
numerous men, women, and children die of rapid
or slow starvation, or of the diseases incidental to
permanently bad conditions of life ; and as there
is nothing to prevent their multiplication before
they are killed off, while, in spite of greater
infant mortality, they increase faster than the
rich ; it seems clear that the struggle for existence in this class can have no appreciable selective influence upon the other 95 per cent. of
the population.

What sort of a sheep breeder would he be who
should content himself with picking out the worst
fifty out of a thousand, leaving them on a
barren common till the weakest starved, and then
letting the survivors go back to mix with the rest ?
And the parallel is too favourable ; since in a
large number of cases, the actual poor and the
convicted criminals are neither the weakest nor
the worst.

In the struggle for the means of enjoyment,
the qualities which ensure success are energy,
industry, intellectual capacity, tenacity of purpose,
and, at least as much sympathy as is necessary
to make a man understand the feelings of his
fellows. Were there none of those artificial arrangements by which fools and knaves are kept at
the top of society instead of sinking to their natural
place at the bottom,[1] the struggle for the means of

---

[1] I have elsewhere lamented the absence from society of

enjoyment would ensure a constant circulation of the human units of the social compound, from the bottom to the top and from the top to the bottom. The survivors of the contest, those who continued to form the great bulk of the polity, would not be those 'fittest' who got to the very top, but the great body of the moderately "fit," whose numbers and superior propagative power, enable them always to swamp the exceptionally endowed minority.

I think it must be obvious to every one, that, whether we consider the internal or the external interests of society, it is desirable they should be in the hands of those who are endowed with the largest share of energy, of industry, of intellectual capacity, of tenacity of purpose, while they are not devoid of sympathetic humanity; and, in so far as the struggle for the means of enjoyment tends to place such men in possession of wealth and influence, it is a process which tends to the good of society. But the process, as we have seen, has no real resemblance to that which adapts living beings to current conditions in the state of nature; nor any to the artificial selection of the horticulturist.

a machinery for facilitating the descent of incapacity. "Administrative Nihilism." *Collected Essays*, vol. i. p. 54.

XV

To return, once more, to the parallel of horti-
culture. In the modern world, the gardening of
men by themselves is practically restricted to the
performance, not of selection, but of that other
function of the gardener, the creation of conditions more favourable than those of the state of
nature; to the end of facilitating the free expansion of the innate faculties of the citizen, so
far as it is consistent with the general good.
And the business of the moral and political
philosopher appears to me to be the ascertainment,
by the same method of observation, experiment,
and ratiocination, as is practised in other kinds
of scientific work, of the course of conduct which
will best conduce to that end.

But, supposing this course of conduct to be
scientifically determined and carefully followed
out, it cannot put an end to the struggle for
existence in the state of nature; and it will not so
much as tend, in any way, to the adaptation of
man to that state. Even should the whole human
race be absorbed in one vast polity, within which
"absolute political justice" reigns, the struggle
for existence with the state of nature outside it,
and the tendency to the return of the struggle
within, in consequence of over-multiplication, will
remain; and, unless men's inheritance from the
ancestors who fought a good fight in the state of

nature, their dose of original sin, is rooted out by
some method at present unrevealed, at any rate
to disbelievers in supernaturalism, every child
born into the world will still bring with him the
instinct of unlimited self-assertion. He will have
to learn the lesson of self-restraint and renunciation. But the practice of self-restraint and renunciation is not happiness, though it may be
something much better.

That man, as a 'political animal,' is susceptible of a vast amount of improvement, by education, by instruction, and by the application of his
intelligence to the adaptation of the conditions
of life to his higher needs, I entertain not the
slightest doubt. But, so long as he remains liable
to error, intellectual or moral; so long as he is
compelled to be perpetually on guard against the
cosmic forces, whose ends are not his ends, without
and within himself; so long as he is haunted by
inexpugnable memories and hopeless aspirations;
so long as the recognition of his intellectual limitations forces him to acknowledge his incapacity to
penetrate the mystery of existence; the prospect
of attaining untroubled happiness, or of a state
which can, even remotely, deserve the title of
perfection, appears to me to be as misleading an
illusion as ever was dangled before the eyes of poor
humanity. And there have been many of them.

That which lies before the human race is a
constant struggle to maintain and improve, in

opposition to the State of Nature, the State of Art
of an organized polity; in which, and by which,
man may develop a worthy civilization, capable
of maintaining and constantly improving itself,
until the evolution of our globe shall have entered
so far upon its downward course that the cosmic
process resumes its sway; and, once more, the
State of Nature prevails over the surface of our
planet.

Note (see p. 30).—It seems the fashion nowadays to ignore
Hartley; though, a century and a half ago, he not only laid
the foundations but built up much of the superstructure of a
true theory of the Evolution of the intellectual and moral
faculties. He speaks of what I have termed the ethical process
as "our Progress from Self-interest to Self-annihilation."
Observations on Man (1749), vol. ii. p. 281.

II

EVOLUTION AND ETHICS

[The Romanes Lecture, 1893]

Soleo enim et in aliena castra transire, non tanquam transfuga sed tanquam explorator. (L. ANNÆI SENECÆ EPIST. II. 4.)

THERE is a delightful child's story, known by the title of " Jack and the Bean-stalk," with which my contemporaries who are present will be familiar. But so many of our grave and reverend juniors have been brought up on severer intellectual diet, and, perhaps, have become acquainted with fairyland only through primers of comparative mythology, that it may be needful to give an outline of the tale. It is a legend of a bean-plant, which grows and grows until it reaches the high heavens and there spreads out into a vast canopy of foliage. The hero, being moved to climb the stalk, discovers that the leafy expanse supports a world composed of the same elements as that below, but yet strangely new ; and his adventures there, on which I may not dwell, must have com-

pletely changed his views of the nature of things;
though the story, not having been composed by,
or for, philosophers, has nothing to say about
views.

My present enterprise has a certain analogy to
that of the daring adventurer. I beg you to
accompany me in an attempt to reach a world
which, to many, is probably strange, by the help
of a bean. It is, as you know, a simple, inert-
looking thing. Yet, if planted under proper conditions, of which sufficient warmth is one of the
most important, it manifests active powers of a
very remarkable kind. A small green seedling
emerges, rises to the surface of the soil, rapidly
increases in size and, at the same time, undergoes
a series of metamorphoses which do not excite our
wonder as much as those which meet us in
legendary history, merely because they are to be
seen every day and all day long.

By insensible steps, the plant builds itself up
into a large and various fabric of root, stem, leaves,
flowers, and fruit, every one moulded within and
without in accordance with an extremely complex
but, at the same time, minutely defined pattern.
In each of these complicated structures, as in their
smallest constituents, there is an immanent energy
which, in harmony with that resident in all the
others, incessantly works towards the maintenance
of the whole and the efficient performance of the
part which it has to play in the economy of nature.

But no sooner has the edifice, reared with such exact elaboration, attained completeness, than it begins to crumble. By degrees, the plant withers and disappears from view, leaving behind more or fewer apparently inert and simple bodies, just like the bean from which it sprang; and, like it, endowed with the potentiality of giving rise to a similar cycle of manifestations.

Neither the poetic nor the scientific imagination is put to much strain in the search after analogies with this process of going forth and, as it were, returning to the starting-point. It may be likened to the ascent and descent of a slung stone, or the course of an arrow along its trajectory. Or we may say that the living energy takes first an upward and then a downward road. Or it may seem preferable to compare the expansion of the germ into the full-grown plant, to the unfolding of a fan, or to the rolling forth and widening of a stream; and thus to arrive at the conception of 'development, or 'evolution.' Here as elsewhere, names are 'noise and smoke'; the important point is to have a clear and adequate conception of the fact signified by a name. And, in this case, the fact is the Sisyphæan process, in the course of which, the living and growing plant passes from the relative simplicity and latent potentiality of the seed to the full epiphany of a highly differentiated type, thence to fall back to simplicity and potentiality.

The value of a strong intellectual grasp of the nature of this process lies in the circumstance that what is true of the bean is true of living things in general. From very low forms up to the highest —in the animal no less than in the vegetable kingdom—the process of life presents the same appearance [1] of cyclical evolution. Nay, we have but to cast our eyes over the rest of the world and cyclical change presents itself on all sides. It meets us in the water that flows to the sea and returns to the springs; in the heavenly bodies that wax and wane, go and return to their places; in the inexorable sequence of the ages of man's life; in that successive rise, apogee, and fall of dynasties and of states which is the most prominent topic of civil history.

As no man fording a swift stream can dip his foot twice into the same water, so no man can, with exactness, affirm of anything in the sensible world that it is.[2] As he utters the words, nay, as he thinks them, the predicate ceases to be applicable; the present has become the past; the 'is' should be 'was.' And the more we learn of the nature of things, the more evident is it that what we call rest is only unperceived activity; that seeming peace is silent but strenuous battle. In every part, at every moment, the state of the cosmos is the expression of a transitory adjustment of contending forces; a scene of strife, in which all the combatants fall in turn. What is true of each part, is true of the whole. Natural knowledge tends more and more to the conclusion that " all the choir of heaven and furniture of the earth " are the transitory forms of parcels of cosmic substance wending along the road of evolution, from nebulous potentiality, through endless growths of sun and planet and satellite; through all varieties of matter; through infinite diversities of life and thought; possibly, through modes of being of which we neither have a conception, nor are competent to form any, back to the indefinable latency from which they arose. Thus the most obvious attribute of the cosmos is its impermanence. It assumes the aspect not so much of a permanent entity as of a changeful process, in which naught endures save the flow of energy and the rational order which pervades it.

We have climbed our bean-stalk and have reached a wonderland in which the common and the familiar become things new and strange. In the exploration of the cosmic process thus typified, the highest intelligence of man finds inexhaustible employment; giants are subdued to our service; and the spiritual affections of the contemplative philosopher are engaged by beauties worthy of eternal constancy.

But there is another aspect of the cosmic process, so perfect as a mechanism, so beautiful as a work of art. Where the cosmopoietic energy works

through sentient beings, there arises, among its
other manifestations, that which we call pain or
suffering. This baleful product of evolution increases in quantity and in intensity, with advancing
grades of animal organization, until it attains its
highest level in man. Further, the consummation is not reached in man, the mere animal; nor
in man, the whole or half savage; but only in
man, the member of an organized polity. And
it is a necessary consequence of his attempt to live
in this way; that is, under those conditions which
are essential to the full development of his noblest
powers.

Man, the animal, in fact, has worked his way
to the headship of the sentient world, and has
become the superb animal which he is, in virtue
of his success in the struggle for existence. The
conditions having been of a certain order, man's
organization has adjusted itself to them better
than that of his competitors in the cosmic strife.
In the case of mankind, the self-assertion, the
unscrupulous seizing upon all that can be grasped,
the tenacious holding of all that can be kept,
which constitute the essence of the struggle for
existence, have answered. For his successful progress, throughout the savage state, man has been
largely indebted to those qualities which he shares
with the ape and the tiger; his exceptional
physical organization; his cunning, his sociability,
his curiosity, and his imitativeness; his ruthless

and ferocious destructiveness when his anger is roused by opposition.

But, in proportion as men have passed from anarchy to social organization, and in proportion as civilization has grown in worth, these deeply ingrained serviceable qualities have become defects. After the mannor of successful persons, civilized man would gladly kick down the ladder by which he has climbed. He would be only too pleased to see 'the ape and tiger die.' But they decline to suit his convenience; and the unwelcome intrusion of these boon companions of his hot youth into the ranged existence of civil life adds pains and griefs, innumerable and immeasurably great, to those which the cosmic process necessarily brings on the mere animal. In fact, civilized man brands all these ape and tiger promptings with the name of sins; he punishes many of the acts which flow from them as crimes; and, in extreme cases, he does his best to put an end to the survival of the fittest of former days by axe and rope.

I have said that civilized man has reached this point; the assertion is perhaps too broad and general; I had better put it that ethical man has attained thereto. The science of ethics professes to furnish us with a reasoned rule of life; to tell us what is right action and why it is so. Whatever differences of opinion may exist among experts, there is a general consensus that the ape and tiger methods of the struggle for existence are not reconcilable with sound ethical principles.

The hero of our story descended the bean-stalk, and came back to the common world, where fare and work were alike hard ; where ugly competitors were much commoner than beautiful princesses ; and where the everlasting battle with self was much less sure to be crowned with victory than a turn-to with a giant. We have done the like. Thousands upon thousands of our fellows, thousands of years ago, have preceded us in finding themselves face to face with the same dread problem of evil. They also have seen that the cosmic process is evolution ; that it is full of wonder, full of beauty, and, at the same time, full of pain. They have sought to discover the bearing of these great facts on ethics ; to find out whether there is, or is not, a sanction for morality in the ways of the cosmos.

Theories of the universe, in which the conception of evolution plays a leading part, were extant at least six centuries before our era. Certain knowledge of them, in the fifth century, reaches us from localities as distant as the valley of the Ganges and the Asiatic coasts of the Ægean. To the early philosophers of Hindostan, no less than to those of Ionia, the salient and characteristic feature of the phenomenal world was its change-

fulness ; the unresting flow of all things, through
birth to visible being and thence to not being, in
which they could discern no sign of a beginning
and for which they saw no prospect of an ending.
It was no less plain to some of these antique fore-
runners of modern philosophy that suffering is the
badge of all the tribe of sentient things ; that it
is no accidental accompaniment, but an essential
constituent of the cosmic process. The energetic
Greek might find fierce joys in a world in which
' strife is father and king ' ; but the old Aryan
spirit was subdued to quietism in the Indian sage ;
the mist of suffering which spread over humanity
hid everything else from his view ; to him life
was one with suffering and suffering with life.

In Hindostan, as in Ionia, a period of relatively
high and tolerably stable civilization had succeeded
long ages of semi-barbarism and struggle. Out of
wealth and security had come leisure and refinement, and, close at their heels, had followed the
malady of thought. To the struggle for bare
existence, which never ends, though it may be
alleviated and partially disguised for a fortunate
few, succeeded the struggle to make existence
intelligible and to bring the order of things into
harmony with the moral sense of man, which also
never ends, but, for the thinking few, becomes
keener with every increase of knowledge and with
every step towards the realization of a worthy
ideal of life.

Two thousand five hundred years ago, the value of civilization was as apparent as it is now; then, as now, it was obvious that only in the garden of an orderly polity can the finest fruits humanity is capable of bearing be produced. But it had also become evident that the blessings of culture were not unmixed. The garden was apt to turn into a hothouse. The stimulation of the senses, the pampering of the emotions, endlessly multiplied the sources of pleasure. The constant widening of the intellectual field indefinitely extended the range of that especially human faculty of looking before and after, which adds to the fleeting present those old and new worlds of the past and the future, wherein men dwell the more the higher their culture. But that very sharpening of the sense and that subtle refinement of emotion, which brought such a wealth of pleasures, were fatally attended by a proportional enlargement of the capacity for suffering; and the divine faculty of imagination, while it created new heavens and new earths, provided them with the corresponding hells of futile regret for the past and morbid anxiety for the future.[3] Finally, the inevitable penalty of over-stimulation, exhaustion, opened the gates of civilization to its great enemy, ennui; the stale and flat weariness when man delights not, nor woman neither; when all things are vanity and vexation; and life seems not worth living except to escape the bore of dying.

Even purely intellectual progress brings about
its revenges. Problems settled in a rough and
ready way by rude men, absorbed in action,
demand renewed attention and show themselves
to be still unread riddles when men have time to
think. The beneficent demon, doubt, whose name
is Legion and who dwells amongst the tombs of
old faiths, enters into mankind and thenceforth
refuses to be cast out. Sacred customs, venerable
dooms of ancestral wisdom, hallowed by tradition
and professing to hold good for all time, are put
to the question. Cultured reflection asks for their
credentials; judges them by its own standards;
finally, gathers those of which it approves into
ethical systems, in which the reasoning is rarely
much more than a decent pretext for the adoption
of foregone conclusions.

One of the oldest and most important elements
in such systems is the conception of justice.
Society is impossible unless those who are associated agree to observe certain rules of conduct
towards one another; its stability depends on the
steadiness with which they abide by that agree-
ment; and, so far as they waver, that mutual
trust which is the bond of society is weakened
or destroyed. Wolves could not hunt in packs
except for the real, though unexpressed, understanding that they should not attack one another
during the chase. The most rudimentary polity
is a pack of men living under the like tacit,

or expressed, understanding; and having made
the very important advance upon wolf society,
that they agree to use the force of the whole body
against individuals who violate it and in favour of
those who observe it. This observance of a common understanding, with the consequent distribution of punishments and rewards according to
accepted rules, received the name of justice, while
the contrary was called injustice. Early ethics
did not take much note of the animus of the
violator of the rules. But civilization could not
advance far, without the establishment of a
capital distinction between the case of involuntary and that of wilful misdeed; between a merely
wrong action and a guilty one. And, with increasing refinement of moral appreciation, the problem
of desert, which arises out of this distinction,
acquired more and more theoretical and practical
importance. If life must be given for life, yet it
was recognized that the unintentional slayer did
not altogether deserve death; and, by a sort of
compromise between the public and the private
conception of justice, a sanctuary was provided
in which he might take refuge from the avenger
of blood.

The idea of justice thus underwent a gradual
sublimation from punishment and reward according to acts, to punishment and reward according to
desert; or, in other words, according to motive.
Righteousness, that is, action from right motive,

not only became synonymous with justice, but the positive constituent of innocence and the very heart of goodness.

Now when the ancient sage, whether Indian or Greek, who had attained to this conception of goodness, looked the world, and especially human life, in the face, he found it as hard as we do to bring the course of evolution into harmony with even the elementary requirements of the ethical ideal of the just and the good.

If there is one thing plainer than another, it is that neither the pleasures nor the pains of life, in the merely animal world, are distributed according to desert; for it is admittedly impossible for the lower orders of sentient beings to deserve either the one or the other. If there is a generalization from the facts of human life which has the assent of thoughtful men in every age and country, it is that the violator of ethical rules constantly escapes the punishment which he deserves; that the wicked flourishes like a green bay tree, while the righteous begs his bread; that the sins of the fathers are visited upon the children; that, in the realm of nature, ignorance is punished just as severely as wilful wrong; and that thousands upon thousands of innocent beings suffer for the crime, or the unintentional trespass, of one.

Greek and Semite and Indian are agreed upon

this subject. The book of Job is at one with the
"Works and Days" and the Buddhist Sutras;
the Psalmist and the Preacher of Israel, with the
Tragic Poets of Greece. What is a more common
motive of the ancient tragedy in fact, than the
unfathomable injustice of the nature of things;
what is more deeply felt to be true than its pre-
sentation of the destruction of the blameless by
the work of his own hands, or by the fatal operation of the sins of others? Surely Œdipus was
pure of heart; it was the natural sequence of
events—the cosmic process—which drove him, in
all innocence, to slay his father and become the
husband of his mother, to the desolation of his
people and his own headlong ruin. Or to step, for
a moment, beyond the chronological limits I have
set myself, what constitutes the sempiternal at-
traction of Hamlet but the appeal to deepest
experience of that history of a no less blameless
dreamer, dragged, in spite of himself, into a world
out of joint; involved in a tangle of crime and
misery, created by one of the prime agents of the
cosmic process as it works in and through man?

Thus, brought before the tribunal of ethics, the
cosmos might well seem to stand condemned.
The conscience of man revolted against the moral
indifference of nature, and the microcosmic atom
should have found the illimitable macrocosm
guilty. But few, or none, ventured to record that
verdict.

In the great Semitic trial of this issue, Job
takes refuge in silence and submission; the Indian
and the Greek, less wise perhaps, attempt to re-
concile the irreconcilable and plead for the defendant. To this end, the Greeks invented Theodicies; while the Indians devised what, in its
ultimate form, must rather be termed a Cos-
modicy. For, though Buddhism recognizes gods
many and lords many, they are products of the
cosmic process; and transitory, however long enduring, manifestations of its eternal activity. In
the doctrine of transmigration, whatever its origin,
Brahminical and Buddhist speculation found,
ready to hand,[4] the means of constructing a
plausible vindication of the ways of the cosmos to
man. If this world is full of pain and sorrow; if
grief and evil fall, like the rain, upon both the
just and the unjust; it is because, like the rain,
they are links in the endless chain of natural
causation by which past, present, and future are
indissolubly connected; and there is no more
injustice in the one case than in the other. Every
sentient being is reaping as it has sown; if not in
this life, then in one or other of the infinite series
of antecedent existences of which it is the latest
term. The present distribution of good and evil
is, therefore, the algebraical sum of accumulated
positive and negative deserts; or, rather, it
depends on the floating balance of the account.
For it was not thought necessary that a complete

settlement should ever take place. Arrears might stand over as a sort of 'hanging gale'; a period of celestial happiness just earned might be succeeded by ages of torment in a hideous nether world, the balance still overdue for some remote ancestral error.[5]

Whether the cosmic process looks any more moral than at first, after such a vindication, may perhaps be questioned. Yet this plea of justification is not less plausible than others; and none but very hasty thinkers will reject it on the ground of inherent absurdity. Like the doctrine of evolution itself, that of transmigration has its roots in the world of reality; and it may claim such support as the great argument from analogy is capable of supplying.

Everyday experience familiarizes us with the facts which are grouped under the name of heredity. Every one of us bears upon him obvious marks of his parentage, perhaps of remoter relationships. More particularly, the sum of tendencies to act in a certain way, which we call "character," is often to be traced through a long series of progenitors and collaterals. So we may justly say that this 'character'—this moral and intellectual essence of a man—does veritably pass over from one fleshly tabernacle to another, and does really transmigrate from generation to generation. In the new-born infant, the character of the stock lies latent, and the Ego is little more than a bundle of potentialities. But, very early, these become actualities; from childhood to age they manifest themselves in dulness or brightness, weakness or strength, viciousness or uprightness; and with each feature modified by confluence with another character, if by nothing else, the character passes on to its incarnation in new bodies.

The Indian philosophers called character, as thus defined, 'karma.' [6] It is this karma which passed from life to life and linked them in the chain of transmigrations; and they held that it is modified in each life, not merely by confluence of parentage, but by its own acts. They were, in fact, strong believers in the theory, so much disputed just at present, of the hereditary transmission of acquired characters. That the manifestation of the tendencies of a character may be greatly facilitated, or impeded, by conditions, of which self-discipline, or the absence of it, are among the most important, is indubitable; but that the character itself is modified in this way is by no means so certain; it is not so sure that the transmitted character of an evil liver is worse, or that of a righteous man better, than that which he received. Indian philosophy, however, did not admit of any doubt on this subject; the belief in the influence of conditions, notably of self-discipline, on the karma was not merely a necessary postulate of its theory of retribution, but it presented the only way of escape from the endless round of transmigrations.

The earlier forms of Indian philosophy agreed with those prevalent in our own times, in supposing the existence of a permanent reality, or 'substance,' beneath the shifting series of phenomena, whether of matter or of mind. The substance of the cosmos was 'Brahma,' that of the individual man 'Atman'; and the latter was separated from the former only, if I may so speak, by its phenomenal envelope, by the casing of sensations, thoughts and desires, pleasures and pains, which make up the illusive phantasmagoria of life. This the ignorant take for reality; their 'Atman' therefore remains eternally imprisoned in delusions, bound by the fetters of desire and scourged by the whip of misery. But the man who has attained enlightenment sees that the apparent reality is mere illusion, or, as was said a couple of thousand years later, that there is nothing good nor bad but thinking makes it so. If the cosmos "is just and of our pleasant vices makes instruments to scourge us," it would seem that the only way to escape from our heritage of evil is to destroy that fountain of desire whence our vices flow; to refuse any longer to be the instruments of the evolutionary process, and withdraw from the struggle for existence. If the karma is modifiable by self-discipline, if its coarser desires, one after another, can be extinguished, the ultimate fundamental desire of self-assertion, or the desire to be, may also be destroyed.[7] Then the bubble of illusion will burst, and the freed individual 'Atman' will lose itself in the universal 'Brahma.'

Such seems to have been the pre-Buddhistic conception of salvation, and of the way to be followed by those who would attain thereto. No more thorough mortification of the flesh has ever been attempted than that achieved by the Indian ascetic anchorite; no later monachism has so nearly succeeded in reducing the human mind to that condition of impassive quasi-somnambulism, which, but for its acknowledged holiness, might run the risk of being confounded with idiocy.

And this salvation, it will be observed, was to be attained through knowledge, and by action based on that knowledge; just as the experimenter, who would obtain a certain physical or chemical result, must have a knowledge of the natural laws involved and the persistent disciplined will adequate to carry out all the various operations required. The supernatural, in our sense of the term, was entirely excluded. There was no external power which could affect the sequence of cause and effect which gives rise to karma; none but the will of the subject of the karma which could put an end to it.

Only one rule of conduct could be based upon the remarkable theory of which I have endeavoured to give a reasoned outline. It was folly to continue

to exist when an overplus of pain was certain;
and the probabilities in favour of the increase of
misery with the prolongation of existence, were
so overwhelming. Slaying the body only made
matters worse; there was nothing for it but to
slay the soul by the voluntary arrest of all its
activities. Property, social ties, family affections,
common companionship, must be abandoned; the
most natural appetites, even that for food, must
be suppressed, or at least minimized; until all
that remained of a man was the impassive,
extenuated, mendicant monk, self-hypnotised
into cataleptic trances, which the deluded mystic
took for foretastes of the final union with
Brahma.

The founder of Buddhism accepted the chief
postulates demanded by his predecessors. But he
was not satisfied with the practical annihilation
involved in merging the individual existence in
the unconditioned—the Atman in Brahma. It
would seem that the admission of the existence of
any substance whatever—even of the tenuity of
that which has neither quality nor energy and of
which no predicate whatever can be asserted—
appeared to him to be a danger and a snare.
Though reduced to a hypostatized negation,
Brahma was not to be trusted; so long as entity
was there, it might conceivably resume the weary
round of evolution, with all its train of immeasur-
able miseries. Gautama got rid of even that

shade of a shadow of permanent existence by a
metaphysical *tour de force* of great interest to the
student of philosophy, seeing that it supplies the
wanting half of Bishop Berkeley's well-known
idealistic argument.

Granting the premises, I am not aware of any
escape from Berkeley's conclusion, that the 'substance' of matter is a metaphysical unknown
quantity, of the existence of which there is no
proof. What Berkeley does not seem to have so
clearly perceived is that the non-existence of a
substance of mind is equally arguable ; and that
the result of the impartial applications of his
reasonings is the reduction of the All to co-
existences and sequences of phenomena, beneath
and beyond which there is nothing cognoscible.
It is a remarkable indication of the subtlety of
Indian speculation that Gautama should have
seen deeper than the greatest of modern idealists ;
though it must be admitted that, if some of
Berkeley's reasonings respecting the nature of
spirit are pushed home, they reach pretty much
the same conclusion.[8]

Accepting the prevalent Brahminical doctrine
that the whole cosmos, celestial, terrestrial, and
infernal, with its population of gods and other
celestial beings, of sentient animals, of Mara and
his devils, is incessantly shifting through recurring
cycles of production and destruction, in each of
which every human being has his transmigratory

representative, Gautama proceeded to eliminate substance altogether; and to reduce the cosmos to a mere flow of sensations, emotions, volitions, and thoughts, devoid of any substratum. As, on the surface of a stream of water, we see ripples and whirlpools, which last for a while and then vanish with the causes that gave rise to them, so what seem individual existences are mere temporary associations of phenomena circling round a centre, "like a dog tied to a post." In the whole universe there is nothing permanent, no eternal substance either of mind or of matter. Personality is a metaphysical fancy; and in very truth, not only we, but all things, in the worlds without end of the cosmic phantasmagoria, are such stuff as dreams are made of.

What then becomes of karma? Karma remains untouched. As the peculiar form of energy we call magnetism may be transmitted from a loadstone to a piece of steel, from the steel to a piece of nickel, as it may be strengthened or weakened by the conditions to which it is subjected while resident in each piece, so it seems to have been conceived that karma might be transmitted from one phenomenal association to another by a sort of induction. However this may be, Gautama doubtless had a better guarantee for the abolition of transmigration, when no wrack of substance, either of Atman or of Brahma, was left behind when, in short, a man had but to dream that he willed not to dream, to put an end to all dreaming.

This end of life's dream is Nirvana. What Nirvana is the learned do not agree. But, since the best original authorities tell us there is neither desire nor activity, nor any possibility of phenomenal reappearance for the sage who has entered Nirvana, it may be safely said of this acme of Buddhistic philosophy—" the rest is silence."[9]

Thus there is no very great practical disagreement between Gautama and his predecessors with respect to the end of action; but it is otherwise as regards the means to that end. With just insight into human nature, Gautama declared extreme ascetic practices to be useless and indeed harmful. The appetites and the passions are not to be abolished by mere mortification of the body; they must, in addition, be attacked on their own ground and conquered by steady cultivation of the mental habits which oppose them; by universal benevolence; by the return of good for evil; by humility; by abstinence from evil thought; in short, by total renunciation of that self-assertion which is the essence of the cosmic process.

Doubtless, it is to these ethical qualities that Buddhism owes its marvellous success.[10] A system which knows no God in the western sense; which denies a soul to man; which counts the belief in immortality a blunder and the hope of it a sin;

which refuses any efficacy to prayer and sacrifice; which bids men look to nothing but their own efforts for salvation; which, in its original purity, knew nothing of vows of obedience, abhorred intolerance, and never sought the aid of the secular arm; yet spread over a considerable moiety of the Old World with marvellous rapidity, and is still, with whatever base admixture of foreign superstitions, the dominant creed of a large fraction of mankind.

Let us now set our faces westwards, towards Asia Minor and Greece and Italy, to view the rise and progress of another philosophy, apparently independent, but no less pervaded by the conception of evolution.[11]

The sages of Miletus were pronounced evolutionists; and, however dark may be some of the sayings of Heracleitus of Ephesus, who was probably a contemporary of Gautama, no better expressions of the essence of the modern doctrine of evolution can be found than are presented by some of his pithy aphorisms and striking metaphors.[12]   Indeed, many of my present auditors must have observed that, more than once, I have borrowed from him in the brief exposition of the theory of evolution with which this discourse commenced.

But when the focus of Greek intellectual activity shifted to Athens, the leading minds concentrated

their attention upon ethical problems. Forsaking
the study of the macrocosm for that of the microcosm, they lost the key to the thought of the
great Ephesian, which, I imagine, is more intelligible to us than it was to Socrates, or to Plato.
Socrates, more especially, set the fashion of a kind
of inverse agnosticism, by teaching that the problems of physics lie beyond the reach of the
human intellect; that the attempt to solve them
is essentially vain; that the one worthy object of
investigation is the problem of ethical life; and
his example was followed by the Cynics and the
later Stoics. Even the comprehensive knowledge
and the penetrating intellect of Aristotle failed to
suggest to him that in holding the eternity of the
world, within its present range of mutation, he was
making a retrogressive step. The scientific heritage of Heracleitus passed into the hands neither
of Plato nor of Aristotle, but into those of Demo-
critus. But the world was not yet ready to
receive the great conceptions of the philosopher of
Abdera. It was reserved for the Stoics to return
to the track marked out by the earlier philosophers; and, professing themselves disciples of
Heracleitus, to develop the idea of evolution
systematically. In doing this, they not only
omitted some characteristic features of their
master's teaching, but they made additions altogether foreign to it. One of the most influential of these importations was the transcen-

dental theism which had come into vogue. The
restless, fiery energy, operating according to law,
out of which all things emerge and into which
they return, in the endless successive cycles of
the great year; which creates and destroys worlds
as a wanton child builds up, and anon levels, sand
castles on the seashore; was metamorphosed into
a material world-soul and decked out with all the
attributes of ideal Divinity; not merely with infinite power and transcendent wisdom, but with
absolute goodness.

The consequences of this step were momentous.
For if the cosmos is the effect of an immanent,
omnipotent, and infinitely beneficent cause, the existence in it of real evil, still less of necessarily
inherent evil, is plainly inadmissible.[13] Yet the
universal experience of mankind testified then, as
now, that, whether we look within us or without
us, evil stares us in the face on all sides; that if
anything is real, pain and sorrow and wrong are
realities.

It would be a new thing in history if *à priori*
philosophers were daunted by the factious opposition of experience; and the Stoics were the last
men to allow themselves to be beaten by mere
facts. 'Give me a doctrine and I will find the
reasons for it,' said Chrysippus. So they perfected, if they did not invent, that ingenious and
plausible form of pleading, the Theodicy; for the
purpose of showing firstly, that there is no such

thing as evil; secondly, that if there is, it is the
necessary correlate of good; and, moreover, that it
is either due to our own fault, or inflicted for our
benefit. Theodicies have been very popular in
their time, and I believe that a numerous, though
somewhat dwarfed, progeny of them still survives.
So far as I know, they are all variations of the
theme set forth in those famous six lines of the
"Essay on Man," in which Pope sums up Boling-
broke's reminiscences of stoical and other speculations of this kind—

> "All nature is but art, unknown to thee;
> All chance, direction which thou canst not see;
> All discord, harmony not understood;
> All partial evil, universal good;
> And spite of pride, in erring reason's spite
> One truth is clear: whatever is is right."

Yet, surely, if there are few more important
truths than those enunciated in the first triad, the
second is open to very grave objections. That
there is a 'soul of good in things evil' is un-
questionable; nor will any wise man deny the
disciplinary value of pain and sorrow. But these
considerations do not help us to see why the immense multitude of irresponsible sentient beings,
which cannot profit by such discipline, should
suffer; nor why, among the endless possibilities
open to omnipotence—that of sinless, happy existence among the rest—the actuality in which sin
and misery abound should be that selected.

Surely it is mere cheap rhetoric to call arguments which have never yet been answered by even the meekest and the least rational of Optimists, suggestions of the pride of reason. As to the concluding aphorism, its fittest place would be as an inscription in letters of mud over the portal of some ' stye of Epicurus ' ; [14] for that is where the logical application of it to practice would land men, with every aspiration stifled and every effort paralyzed. Why try to set right what is right already? Why strive to improve the best of all possible worlds? Let us eat and drink, for as to-day all is right, so to-morrow all will be.

But the attempt of the Stoics to blind themselves to the reality of evil, as a necessary concomitant of the cosmic process, had less success than that of the Indian philosophers to exclude the reality of good from their purview. Unfortunately, it is much easier to shut one's eyes to good than to evil. Pain and sorrow knock at our doors more loudly than pleasure and happiness ; and the prints of their heavy footsteps are less easily effaced. Before the grim realities of practical life the pleasant fictions of optimism vanished. If this were the best of all possible worlds, it nevertheless proved itself a very inconvenient habitation for the ideal sage.

The stoical summary of the whole duty of man, ' Live according to nature,' would seem to imply that the cosmic process is an exemplar for human

conduct. Ethics would thus become applied
Natural History. In fact, a confused employment
of the maxim, in this sense, has done immeasur-
able mischief in later times. It has furnished an
axiomatic foundation for the philosophy of philo-
sophasters and for the moralizing of sentimentalists.
But the Stoics were, at bottom, not merely noble,
but sane, men; and if we look closely into what
they really meant by this ill-used phrase, it will
be found to present no justification for the mischievous conclusions that have been deduced
from it.

In the language of the Stoa, ' Nature ' was a
word of many meanings. There was the ' Nature '
of the cosmos and the ' Nature ' of man. In the
latter, the animal 'nature,' which man shares
with a moiety of the living part of the cosmos, was
distinguished from a higher 'nature.' Even in
this higher nature there were grades of rank.
The logical faculty is an instrument which may be
turned to account for any purpose. The passions
and the emotions are so closely tied to the lower
nature that they may be considered to be patho-
logical, rather than normal, phenomena. The one
supreme, hegemonic, faculty, which constitutes the
essential ' nature ' of man, is most nearly represented by that which, in the language of a later
philosophy, has been called the pure reason. It is
this ' nature ' which holds up the ideal of the
supreme good and demands absolute submission of

the will to its behests.   It is this which commands
all men to love one another, to return good for evil, to
regard one another as fellow-citizens of one great
state.   Indeed, seeing that the progress towards
perfection of a civilized state, or polity, depends
on the obedience of its members to these commands, the Stoics sometimes termed the pure
reason the 'political' nature.   Unfortunately,
the sense of the adjective has undergone so much
modification, that the application of it to that
which commands the sacrifice of self to the
common good would now sound almost grotesque.[15]

But what part is played by the theory of evolution in this view of ethics ?   So far as I can
discern, the ethical system of the Stoics, which is
essentially intuitive, and reverences the categorical
imperative as strongly as that of any later
moralists, might have been just what it was if they
had held any other theory ; whether that of special
creation, on the one side, or that of the eternal
existence of the present order, on the other.[16]   To
the Stoic, the cosmos had no importance for the
conscience, except in so far as he chose to think
it a pedagogue to virtue.   The pertinacious optimism of our philosophers hid from them the actual
state of the case.   It prevented them from seeing
that cosmic nature is no school of virtue, but the
headquarters of the enemy of ethical nature.
The logic of facts was necessary to convince them

that the cosmos works through the lower nature of man, not for righteousness, but against it. And it finally drove them to confess that the existence of their ideal " wise man " was incompatible with the nature of things ; that even a passable approximation to that ideal was to be attained only at the cost of renunciation of the world and mortification, not merely of the flesh, but of all human affections. The state of perfection was that ' apatheia '[17] in which desire, though it may still be felt, is powerless to move the will, reduced to the sole function of executing the commands of pure reason. Even this residuum of activity was to be regarded as a temporary loan, as an efflux of the divine world-pervading spirit, chafing at its imprisonment in the flesh, until such time as death enabled it to return to its source in the all-pervading logos.

I find it difficult to discover any very great difference between Apatheia and Nirvana, except that stoical speculation agrees with pre-Buddhistic philosophy, rather than with the teachings of Gautama, in so far as it postulates a permanent substance equivalent to ' Brahma ' and ' Atman '; and that, in stoical practice, the adoption of the life of the mendicant cynic was held to be more a counsel of perfection than an indispensable condition of the higher life.

Thus the extremes touch.   Greek thought and

Indian thought set out from ground common to
both, diverge widely, develop under very different
physical and moral conditions, and finally converge
to practically the same end.

The Vedas and the Homeric epos set before us
a world of rich and vigorous life, full of joyous
fighting men

> That ever with a frolic welcome took
> The thunder and the sunshine . . . .

and who were ready to brave the very Gods themselves when their blood was up. A few centuries
pass away, and under the influence of civilization
the descendants of these men are ' sicklied o'er
with the pale cast of thought '—frank pessimists,
or, at best, make-believe optimists. The courage
of the warlike stock may be as hardly tried as
before, perhaps more hardly, but the enemy is
self. The hero has become a monk. The man of
action is replaced by the quietist, whose highest
aspiration is to be the passive instrument of the
divine Reason. By the Tiber, as by the Ganges,
ethical man admits that the cosmos is too strong
for him ; and, destroying every bond which ties
him to it by ascetic discipline, he seeks salvation
in absolute renunciation.[18]

Modern thought is making a fresh start from
the base whence Indian and Greek philosophy set
out ; and, the human mind being very much what

it was six-and-twenty centuries ago, there is no ground for wonder if it presents indications of a tendency to move along the old lines to the same results.

We are more than sufficiently familiar with modern pessimism, at least as a speculation; for I cannot call to mind that any of its present votaries have sealed their faith by assuming the rags and the bowl of the mendicant Bhikku, or the cloak and the wallet of the Cynic. The obstacles placed in the way of sturdy vagrancy by an unphilosophical police have, perhaps, proved too formidable for philosophical consistency. We also know modern speculative optimism, with its perfectibility of the species, reign of peace, and lion and lamb transformation scenes; but one does not hear so much of it as one did forty years ago; indeed, I imagine it is to be met with more commonly at the tables of the healthy and wealthy, than in the congregations of the wise. The majority of us, I apprehend, profess neither pessimism nor optimism. We hold that the world is neither so good, nor so bad, as it conceivably might be; and, as most of us have reason, now and again, to discover that it can be. Those who have failed to experience the joys that make life worth living are, probably, in as small a minority as those who have never known the griefs that rob existence of its savour and turn its richest fruits into mere dust and ashes.

Further, I think I do not err in assuming that, however diverse their views on philosophical and religious matters, most men are agreed that the proportion of good and evil in life may be very sensibly affected by human action. I never heard anybody doubt that the evil may be thus increased, or diminished; and it would seem to follow that good must be similarly susceptible of addition or subtraction. Finally, to my knowledge, nobody professes to doubt that, so far forth as we possess a power of bettering things, it is our paramount duty to use it and to train all our intellect and energy to this supreme service of our kind.

Hence the pressing interest of the question, to what extent modern progress in natural knowledge, and, more especially, the general outcome of that progress in the doctrine of evolution, is competent to help us in the great work of helping one another?

The propounders of what are called the " ethics of evolution," when the ' evolution of ethics ' would usually better express the object of their speculations, adduce a number of more or less interesting facts and more or less sound arguments, in favour of the origin of the moral sentiments, in the same way as other natural phenomena, by a process of evolution. I have little doubt, for my own part, that they are on the right track; but as the immoral sentiments have no less been evolved, there is, so far, as much natural sanction for the one as the other. The thief and the murderer follow nature just as much as the philanthropist. Cosmic evolution may teach us how the good and the evil tendencies of man may have come about; but, in itself, it is incompetent to furnish any better reason why what we call good is preferable to what we call evil than we had before. Some day, I doubt not, we shall arrive at an understanding of the evolution of the æsthetic faculty; but all the understanding in the world will neither increase nor diminish the force of the intuition that this is beautiful and that is ugly.

There is another fallacy which appears to me to pervade the so-called " ethics of evolution." It is the notion that because, on the whole, animals and plants have advanced in perfection of organization by means of the struggle for existence and the consequent ' survival of the fittest '; therefore men in society, men as ethical beings, must look to the same process to help them towards perfection. I suspect that this fallacy has arisen out of the unfortunate ambiguity of the phrase ' survival of the fittest.' ' Fittest ' has a connotation of ' best '; and about ' best ' there hangs a moral flavour. In cosmic nature, however, what is ' fittest ' depends upon the conditions. Long since,[19] I ventured to point out that if our hemisphere were to cool again, the survival of the fittest might bring about, in the vegetable kingdom, a population of more and more stunted and humbler and humbler organisms, until the 'fittest' that survived might be nothing but lichens, diatoms, and such microscopic organisms as those which give red snow its colour; while, if it became hotter, the pleasant valleys of the Thames and Isis might be uninhabitable by any animated beings save those that flourish in a tropical jungle. They, as the fittest, the best adapted to the changed conditions, would survive.

Men in society are undoubtedly subject to the cosmic process. As among other animals, multiplication goes on without cessation, and involves severe competition for the means of support. The struggle for existence tends to eliminate those less fitted to adapt themselves to the circumstances of their existence. The strongest, the most self-assertive, tend to tread down the weaker. But the influence of the cosmic process on the evolution of society is the greater the more rudimentary its civilization. Social progress means a checking of the cosmic process at every step and the substitution for it of another, which may be called the ethical process; the end of which is not the survival of those who may happen to be the fittest, in respect of the whole of the conditions which obtain, but of those who are ethically the best.[20]

As I have already urged, the practice of that which is ethically best—what we call goodness or virtue—involves a course of conduct which, in all

respects, is opposed to that which leads to success
in the cosmic struggle for existence.   In place of
ruthless self-assertion it demands self-restraint;
in place of thrusting aside, or treading down, all
competitors, it requires that the individual shall
not merely respect, but shall help his fellows; its
influence is directed, not so much to the survival
of the fittest, as to the fitting of as many as possible to survive.   It repudiates the gladiatorial
theory of existence.   It demands that each man
who enters into the enjoyment of the advantages
of a polity shall be mindful of his debt to those
who have laboriously constructed it; and shall
take heed that no act of his weakens the fabric in
which he has been permitted to live.   Laws and
moral precepts are directed to the end of curbing
the cosmic process and reminding the individual
of his duty to the community, to the protection
and influence of which he owes, if not existence
itself, at least the life of something better than a
brutal savage.

It is from neglect of these plain considerations
that the fanatical individualism [21] of our time
attempts to apply the analogy of cosmic nature to
society.   Once more we have a misapplication of
the stoical injunction to follow nature; the duties
of the individual to the state are forgotten, and his
tendencies to self-assertion are dignified by the
name of rights.   It is seriously debated whether
the members of a community are justified in

using their combined strength to constrain one of
their number to contribute his share to the maintenance of it; or even to prevent him from doing
his best to destroy it. The struggle for existence,
which has done such admirable work in cosmic
nature, must, it appears, be equally beneficent in
the ethical sphere. Yet if that which I have insisted upon is true; if the cosmic process has no
sort of relation to moral ends; if the imitation of
it by man is inconsistent with the first principles
of ethics; what becomes of this surprising theory?

Let us understand, once for all, that the ethical
progress of society depends, not on imitating the
cosmic process, still less in running away from it,
but in combating it. It may seem an audacious
proposal thus to pit the microcosm against the
macrocosm and to set man to subdue nature to his
higher ends; but I venture to think that the
great intellectual difference between the ancient
times with which we have been occupied and our
day, lies in the solid foundation we have acquired
for the hope that such an enterprise may meet
with a certain measure of success.

The history of civilization details the steps by
which men have succeeded in building up an
artificial world within the cosmos. Fragile reed
as he may be, man, as Pascal says, is a thinking
reed:[22] there lies within him a fund of energy,
operating intelligently and so far akin to that
which pervades the universe, that it is competent

G 2

to influence and modify the cosmic process. In
virtue of his intelligence, the dwarf bends the
Titan to his will. In every family, in every polity
that has been established, the cosmic process in
man has been restrained and otherwise modified
by law and custom ; in surrounding nature, it has
been similarly influenced by the art of the shepherd, the agriculturist, the artisan. As civilization
has advanced, so has the extent of this interference increased ; until the organized and highly
developed sciences and arts of the present day
have endowed man with a command over the
course of non-human nature greater than that
once attributed to the magicians. The most im-
pressive, I might say startling, of these changes
have been brought about in the course of the last
two centuries ; while a right comprehension of the
process of life and of the means of influencing
its manifestations is only just dawning upon us.
We do not yet see our way beyond generalities;
and we are befogged by the obtrusion of false
analogies and crude anticipations. But Astronomy, Physics, Chemistry, have all had to pass
through similar phases, before they reached the
stage at which their influence became an important factor in human affairs. Physiology, Psycho-
logy, Ethics, Political Science, must submit to the
same ordeal. Yet it seems to me irrational to
doubt that, at no distant period, they will work as
great a revolution in the sphere of practice.

The theory of evolution encourages no millennial anticipations. If, for millions of years, our globe has taken the upward road, yet, some time, the summit will be reached and the downward route will be commenced. The most daring imagination will hardly venture upon the suggestion that the power and the intelligence of man can ever arrest the procession of the great year.

Moreover, the cosmic nature born with us and, to a large extent, necessary for our maintenance, is the outcome of millions of years of severe training, and it would be folly to imagine that a few centuries will suffice to subdue its masterfulness to purely ethical ends. Ethical nature may count upon having to reckon with a tenacious and powerful enemy as long as the world lasts. But, on the other hand, I see no limit to the extent to which intelligence and will, guided by sound principles of investigation, and organized in common effort, may modify the conditions of existence, for a period longer than that now covered by history. And much may be done to change the nature of man himself.[23] The intelligence which has converted the brother of the wolf into the faithful guardian of the flock ought to be able to do something towards curbing the instincts of savagery in civilized men.

But if we may permit ourselves a larger hope of abatement of the essential evil of the world than was possible to those who, in the infancy of exact knowledge, faced the problem of existence more than a score of centuries ago, I deem it an essential condition of the realization of that hope that we should cast aside the notion that the escape from pain and sorrow is the proper object of life.

We have long since emerged from the heroic childhood of our race, when good and evil could be met with the same 'frolic welcome'; the attempts to escape from evil, whether Indian or Greek, have ended in flight from the battle-field; it remains to us to throw aside the youthful over-confidence and the no less youthful discouragement of nonage. We are grown men, and must play the man

strong in will

To strive, to seek, to find, and not to yield,

cherishing the good that falls in our way, and bearing the evil, in and around us, with stout hearts set on diminishing it. So far, we all may strive in one faith towards one hope:

It may be that the gulfs will. wash us down,
It may be we shall touch the Happy Isles,

. . . . but something ere the end,
Some work of noble note may yet be done. (24)

NOTES

Note 1 (p. 49).

I HAVE been careful to speak of the "appearance"
of cyclical evolution presented by living things; for,
on critical examination, it will be found that the
course of vegetable and of animal life is not exactly
represented by the figure of a cycle which returns
into itself. What actually happens, in all but the
lowest organisms, is that one part of the growing
germ (A) gives rise to tissues and organs; while
another part (B) remains in its primitive condition,
or is but slightly modified. The moiety A becomes
the body of the adult and, sooner or later, perishes,
while portions of the moiety B are detached and, as
offspring, continue the life of the species. Thus, if
we trace back an organism along the direct line of
descent from its remotest ancestor, B, as a whole,
has never suffered death; portions of it, only, have
been cast off and died in each individual offspring.

Everybody is familiar with the way in which the
"suckers" of a strawberry plant behave. A thin
cylinder of living tissue keeps on growing at its free
end, until it attains a considerable length. At

successive intervals, it develops buds which grow into strawberry plants ; and these become independent by the death of the parts of the sucker which connect them. The rest of the sucker, however, may go on living and growing indefinitely, and, circumstances remaining favourable, there is no obvious reason why it should ever die. The living substance B, in a manner, answers to the sucker. If we could restore the continuity which was once possessed by the portions of B, contained in all the individuals of a direct line of descent, they would form a sucker, or *stolon*, on which these individuals would be strung, and which would never have wholly died.

A species remains unchanged so long as the potentiality of development resident in B remains unaltered ; so long, *e.g.*, as the buds of the strawberry sucker tend to become typical strawberry plants. In the case of the progressive evolution of a species, the developmental potentiality of B becomes of a higher and higher order. In retrogressive evolution, the contrary would be the case. The phenomena of atavism seem to show that retrogressive evolution, that is, the return of a species to one or other of its earlier forms, is a possibility to be reckoned with. The simplification of structure, which is so common in the parasitic members of a group, however, does not properly come under this head. The worm-like, limbless *Lernœa* has no resemblance to any of the stages of development of the many-limbed active animals of the group to which it belongs.

Note 2 (p. 49)

Heracleitus says, Ποταμῷ γὰρ οὐκ ἔστι δὶς ἐμβῆναι τῷ αὐτῷ; but, to be strictly accurate, the river remains, though the water of which it is composed changes—just as a man retains his identity though the whole substance of his body is constantly shifting. This is put very well by Seneca (Ep. lvii. i. 20, Ed. Ruhkopf): "Corpora nostra rapiuntur fluminum more, quidquid vides currit cum tempore; nihil ex his quæ videmus manet. Ego ipse dum loquor mutari ista, mutatus sum. Hoc est quod ait Heraclitus 'In idem flumen bis non descendimus.' Manet idem fluminis nomen, aqua transmissa est. Hoc in amne manifestius est quam in homine, sed nos quoque non minus velox cursus prætervehit."

Note 3 (p. 55).

"Multa bona nostra nobis nocent, timoris enim tormentum memoria reducit, providentia anticipat. Nemo tantum præsentibus miser est." (Seneca, Ed. v. 7.)

Among the many wise and weighty aphorisms of the Roman Bacon, few sound the realities of life more deeply than "Multa bona nostra nobis nocent." If there is a soul of good in things evil, it is at least equally true that there is a soul of evil in things good: for things, like men, have "les défauts de leurs qualités." It is one of the last lessons one learns from experience, but not the least important, that a

heavy tax is levied upon all forms of success; and
that failure is one of the commonest disguises
assumed by blessings.

Note 4 (p. 60).

"There is within the body of every man a soul
which, at the death of the body, flies away from it
like a bird out of a cage, and enters upon a new
life . . . either in one of the heavens or one of the
hells or on this earth. The only exception is the
rare case of a man having in this life acquired a
true knowledge of God. According to the pre-
Buddhistic theory, the soul of such a man goes along
the path of the Gods to God, and, being united with
Him, enters upon an immortal life in which his
individuality is not extinguished. In the latter theory,
his soul is directly absorbed into the Great Soul, is
lost in it, and has no longer any independent existence.
The souls of all other men enter, after the death of
the body, upon a new existence in one or other of
the many different modes of being. If in heaven or
hell, the soul itself becomes a god or demon without
entering a body; all superhuman beings, save the
great gods, being looked upon as not eternal, but
merely temporary creatures. If the soul returns to
earth it may or may not enter a new body; and this
either of a human being, an animal, a plant, or even
a material object. For all these are possessed of
souls, and there is no essential difference between
these souls and the souls of men—all being alike
mere sparks of the Great Spirit, who is the only real

existence." (Rhys Davids, *Hibbert Lectures*, 1881, p. 83.)

For what I have said about Indian Philosophy, I am particularly indebted to the luminous exposition of primitive Buddhism and its relations to earlier Hindu thought, which is given by Prof. Rhys Davids in his remarkable *Hibbert Lectures* for 1881, and *Buddhism* (1890). The only apology I can offer for the freedom with which I have borrowed from him in these notes, is my desire to leave no doubt as to my indebtedness. I have also found Dr. Oldenberg's *Buddha* (Ed. 2, 1890) very helpful. The origin of the theory of transmigration stated in the above extract is an unsolved problem. That it differs widely from the Egyptian metempsychosis is clear. In fact, since men usually people the other world with phantoms of this, the Egyptian doctrine would seem to presuppose the Indian as a more archaic belief.

Prof. Rhys Davids has fully insisted upon the ethical importance of the transmigration theory. " One of the latest speculations now being put forward among ourselves would seek to explain each man's character, and even his outward condition in life, by the character he inherited from his ancestors, a character gradually formed during a practically endless series of past existences, modified only by the conditions into which he was born, those very conditions being also, in like manner, the last result of a practically endless series of past causes. Gotama's speculation might be stated in the same words. But it attempted also to explain, in a way different from

that which would be adopted by the exponents of
the modern theory, that strange problem which it
is also the motive of the wonderful drama of the
book of Job to explain—the fact that the actual
distribution here of good fortune, or misery, is
entirely independent of the moral qualities which
men call good or bad. We cannot wonder that a
teacher, whose whole system was so essentially an
ethical reformation, should have felt it incumbent
upon him to seek an explanation of this apparent
injustice. And all the more so, since the belief he
had inherited, the theory of the transmigration of
souls, had provided a solution perfectly sufficient to
any one who could accept that belief." (*Hibbert
Lectures*, p. 93.) I should venture to suggest the
substitution of 'largely' for 'entirely' in the foregoing passage. Whether a ship makes a good or a
bad voyage is largely independent of the conduct of
the captain, but it is largely affected by that conduct. Though powerless before a hurricane he may
weather many a bad gale.

Note 5 (p. 61).

The outward condition of the soul is, in each new
birth, determined by its actions in a previous birth;
but by each action in succession, and not by the
balance struck after the evil has been reckoned off
against the good. A good man who has once uttered
a slander may spend a hundred thousand years as a
god, in consequence of his goodness, and when the
power of his good actions is exhausted, may be born

as a dumb man on account of his transgression ; and
a robber who has once done an act of mercy, may
come to life in a king's body as the result of his
virtue, and then suffer torments for ages in hell or
as a ghost without a body, or be re-born many times
as a slave or an outcast, in consequence of his evil
life.

"There is no escape, according to this theory, from
the result of any act ; though it is only the consequences of its own acts that each soul has to endure.
The force has been set in motion by itself and can
never stop ; and its effect can never be foretold. If evil,
it can never be modified or prevented, for it depends
on a cause already completed, that is now for ever
beyond the soul's control. There is even no continuing
consciousness, no memory of the past that could guide
the soul to any knowledge of its fate. The only
advantage open to it is to add in this life to the sum
of its good actions, that it may bear fruit with the
rest. And even this can only happen in some future
life under essentially the same conditions as the present one : subject, like the present one, to old age,
decay, and death ; and affording opportunity, like the
present one, for the commission of errors, ignorances,
or sins, which in their turn must inevitably produce
their due effect of sickness, disability, or woe. Thus
is the soul tossed about from life to life, from billow
to billow in the great ocean of transmigration. And
there is no escape save for the very few, who, during
their birth as men, attain to a right knowledge of the
Great Spirit : and thus enter into immortality, or, as
the later philosophers taught, are absorbed into the

Divine Essence." (Rhys Davids, *Hibbert Lectures*, pp. 85, 86.)

The state after death thus imagined by the Hindu philosophers has a certain analogy to the purgatory of the Roman Church ; except that escape from it is dependent, not on a divine decree modified, it may be, by sacerdotal or saintly intercession, but by the acts of the individual himself ; and that while ultimate emergence into heavenly bliss of the good, or well-prayed for, Catholic is professedly assured, the chances in favour of the attainment of absorption, or of Nirvana, by any individual Hindu are extremely small.

Note 6 (p. 62).

" That part of the then prevalent transmigration theory which could not be proved false seemed to meet a deeply felt necessity, seemed to supply a moral cause which would explain the unequal distribution here of happinesss or woe, so utterly inconsistent with the present characters of men." Gautama " still therefore talked of men's previous existence, but by no means in the way that he is generally represented to have done." What he taught was "the transmigration of character." He held that after the death of any being, whether human or not, there survived nothing at all but that being's ' Karma,' the result, that is, of its mental and bodily actions. Every individual, whether human or divine, was the last inheritor and the last result of the Karma of a long series of past individuals—a series so long that its beginning is beyond the reach of calculation, and its end will be coincident with the destruction of the world." (Rhys Davids, *Hibbert Lectures*, p. 92.)

In the theory of evolution, the tendency of a germ to develop according to a certain specific type, *e.g.* of the kidney bean seed to grow into a plant having all the characters of *Phaseolus vulgaris*, is its 'Karma.' It is the "last inheritor and the last result" of all the conditions that have affected a line of ancestry which goes back for many millions of years to the time when life first appeared on the earth. The moiety B of the substance of the bean plant (see *Note* 1) is the last link in a once continuous chain extending from the primitive living substance : and the characters of the successive species to which it has given rise are the manifestations of its gradually modified Karma. As Prof. Rhys Davids aptly says, the snowdrop "is a snowdrop and not an oak, and just that kind of snowdrop, because it is the outcome of the Karma of an endless series of past existences." (*Hibbert Lectures*, p. 114.)

Note 7 (p. 64).

"It is interesting to notice that the very point which is the weakness of the theory—the supposed concentration of the effect of the Karma in one new being—presented itself to the early Buddhists themselves as a difficulty. They avoided it, partly by explaining that it was a particular thirst in the creature dying (a craving, Tanhā, which plays otherwise a great part in the Buddhist theory) which actually caused the birth of the new individual who was to inherit the Karma of the former one. But, how this took place, how the craving desire produced this effect, was acknowledged to be a mystery patent only to a Buddha." (Rhys Davids, *Hibbert Lectures*, p. 95.)

Among the many parallelisms of Stoicism and Buddhism, it is curious to find one for this Tanhā, 'thirst,' or 'craving desire' for life. Seneca writes (Epist. lxxvi. 18): "Si enim ullum aliud est bonum quam honestum, sequetur nos *aviditas vitæ* aviditas rerum vitam instruentium: quod est intolerabile infinitum, vagum."

Note 8 (p. 66).

"The distinguishing characteristic of Buddhism was that it started a new line, that it looked upon the deepest questions men have to solve from an entirely different standpoint. It swept away from the field of its vision the whole of the great soul-theory which had hitherto so completely filled and dominated the minds of the superstitious and the thoughtful alike. For the first time in the history of the world, it proclaimed a salvation which each man could gain for himself and by himself, in this world, during this life, without any the least reference to God, or to Gods, either great or small. Like the Upanishads, it placed the first importance on knowledge; but it was no longer a knowledge of God, it was a clear perception of the real nature, as they supposed it to be, of men and things. And it added to the necessity of knowledge, the necessity of purity, of courtesy, of uprightness, of peace and of a universal love far reaching, grown great and beyond measure." (Rhys Davids, *Hibbert Lectures*, p. 29.)

The contemporary Greek philosophy takes an analogous direction. According to Heracleitus, the universe was made neither by Gods nor men ; but, from all eternity, has been, and to all eternity, will be, immortal fire, glowing and fading in due measure. (Mullach, *Heracliti Fragmenta*, 27.) And the part assigned by his successors, the Stoics, to the knowledge and the volition of the ' wise man' made their Divinity (for logical thinkers) a subject for compliments, rather than a power to be reckoned with. In Hindu speculation the ' Arahat,' still more the ' Buddha,' becomes the superior of Brahma ; the stoical ' wise man' is, at least, the equal of Zeus.

Berkeley affirms over and over again that no idea can be formed of a soul or spirit—" If any man shall doubt of the truth of what is here delivered, let him but reflect and try if he can form any idea of power or active being ; and whether he hath ideas of two principal powers marked by the names of *will* and *understanding* distinct from each other, as well as from a third idea of substance or being in general, with a relative notion of its supporting or being the subject of the aforesaid power, which is signified by the name *soul* or *spirit*. This is what some hold : but, so far as I can see, the words *will, soul, spirit,* do not stand for different ideas or, in truth, for any idea at all, but for something which is very different from ideas, and which, being an agent, cannot be like unto or represented by any idea whatever [though it must be owned at the same time, that we have some notion of soul, spirit, and the operations of the mind, such as willing, loving, hating, inasmuch as we know or understand the meaning of these words "]. (*The Principles of Human Knowledge*, lxxvi. See also §§ lxxxix., cxxxv., cxlv.)

It is open to discussion, I think, whether it is possible to have 'some notion' of that of which we can form no 'idea.'

Berkeley attaches several predicates to the " perceiving active being mind, spirit, soul or myself " (Parts I. II.) It is said, for example, to be "indivisible, incorporeal, unextended, and incorruptible." The predicate indivisible, though negative in form, has highly positive consequences. For, if 'perceiving active being' is strictly indivisible, man's soul must be one with the Divine spirit : which is good Hindu or Stoical doctrine, but hardly orthodox Christian philosophy. If, on the other hand, the 'substance' of active perceiving 'being' is actually divided into the one Divine and innumerable human entities, how can the predicate 'indivisible' be rigorously applicable to it ?

Taking the words cited, as they stand, they amount to the denial of the possibility of any knowledge of substance. 'Matter' having been resolved into mere affections of 'spirit,' 'spirit' melts away into an admittedly inconceivable and unknowable hypostasis of thought and power—consequently the existence of anything in the universe beyond a flow of phenomena is a purely hypothetical assumption. Indeed a pyrrhonist might raise the objection that if 'esse' is 'percipi' spirit itself can have no existence except as a perception, hypostatized into a 'self,' or as a perception of some other spirit. In the former case, objective reality vanishes; in the latter, there would seem to be the need of an infinite series of spirits each perceiving the others.

It is curious to observe how very closely the phraseology of Berkeley sometimes approaches that of the Stoics: thus (cxlviii.) "It seems to be a *general pretence of the unthinking* herd that *they cannot see God*......But, alas, we need only open our eyes to see the Sovereign Lord of all things with a more full and clear view, than we do any of our fellow-creatures. ....we do at all times and in all places perceive manifest tokens of the Divinity: everything we see, hear, feel, or any wise perceive by sense, being a sign or effect of the power of God" ......cxlix. "It is therefore plain, that *nothing can be more evident* to any one that is capable of the least reflection, *than the existence of God*, or a spirit who is intimately present to our minds, producing in them all that variety of ideas or sensations which continually affect us, on whom we have an absolute and entire dependence, in short, *in whom we live and move and have our being*." cl. [But you will say hath Nature no share in the production of natural things, and must they be all ascribed to the immediate and sole operation of God?......if by *Nature* is meant some

H 2

being distinct from God, as well as from the laws of
nature and things perceived by sense, I must confess
that word is to me an empty sound, without any
intelligible meaning annexed to it.] Nature in this
acceptation is a vain *Chimæra* introduced by those
heathens, who had not just notions of the omni-
presence and infinite perfection of God."

Compare Seneca (*De Beneficiis*, iv. 7):

" Natura, inquit, hæc mihi præstat. Non intelligis
te, quum hoc dicis, mutare Nomen Deo? Quid enim
est aliud Natura quam Deus, et divina ratio, toti
mundo et partibus ejus inserta? Quoties voles tibi
licet aliter hunc auctorem rerum nostrarum com-
pellare, et Jovem illum optimum et maximum rite
dices, et tonantem, et statorem : qui non, ut historici
tradiderunt, ex eo quod post votum susceptum acies
Romanorum fugientum stetit, sed quod stant beneficio
ejus omnia, stator, stabilitorque est : hunc eundem et
fatum si dixeris, non mentieris, nam quum fatum
nihil aliud est, quam series implexa causarum, ille est
prima omnium causa, ea qua cæteræ pendent." It
would appear, therefore, that the good Bishop is
somewhat hard upon the ' heathen,' of whose words
his own might be a paraphrase.

There is yet another direction in which Berkeley's
philosophy, I will not say agrees with Gautama's, but
at any rate helps to make a fundamental dogma of
Buddhism intelligible.

" I find I can excite ideas in my mind at pleasure,
and vary and shift the scene as often as I think fit.
It is no more than willing, and straightway this or
that idea arises in my fancy : and by the same power,

it is obliterated, and makes way for another. This making and unmaking of ideas doth very properly denominate the mind active. This much is certain and grounded on experience. . . ." (*Principles,* xxviii.)

A good many of us, I fancy, have reason to think that experience tells them very much the contrary; and are painfully familiar with the obsession of the mind by ideas which cannot be obliterated by any effort of the will and steadily refuse to make way for others. But what I desire to point out is that if Gautama was equally confident that he could 'make and unmake' ideas—then, since he had resolved self into a group of ideal phantoms—the possibility of abolishing self by volition naturally followed.

Note 9 (p. 68).

According to Buddhism, the relation of one life to the next is merely that borne by the flame of one lamp to the flame of another lamp which is set alight by it. To the 'Arahat' or adept "no outward form, no compound thing, no creature, no creator, no existence of any kind, must appear to be other than a temporary collocation of its component parts, fated inevitably to be dissolved."—(Rhys Davids, *Hibbert Lectures,* p. 211.)

The self is nothing but a group of phenomena held together by the desire of life; when that desire shall have ceased, "the Karma of that particular chain of lives will cease to influence any longer any distinct individual, and there will be no more birth; for birth, decay, and death, grief, lamentation, and despair will have come, so far as regards that chain of lives, for ever to an end."

The state of mind of the Arahat in which the desire of life has ceased is Nirvana. Dr. Oldenberg has very acutely and patiently considered the various interpretations which have been attached to ' Nirvana ' in the work to which I have referred (pp. 285 *et seq.*). The result of his and other discussions of the question may I think be briefly stated thus :

1. Logical deduction from the predicates attached to the term ' Nirvana ' strips it of all reality, conceivability, or perceivability, whether by Gods or men. For all practical purposes, therefore, it comes to exactly the same thing as annihilation.

2. But it is not annihilation in the ordinary sense, inasmuch as it could take place in the living Arahat or Buddha.

3. And, since, for the faithful Buddhist, that which was abolished in the Arahat was the possibility of further pain, sorrow, or sin ; and that which was attained was perfect peace ; his mind directed itself exclusively to this joyful consummation, and personified the negation of all conceivable existence and of all pain into a positive bliss. This was all the more easy, as Gautama refused to give any dogmatic definition of Nirvana. There is something analogous in the way in which people commonly talk of the ' happy release ' of a man who has been long suffering from mortal disease. According to their own views, it must always be extremely doubtful whether the man will be any happier after the ' release ' than

before. But they do not choose to look at the matter in this light.

The popular notion that, with practical, if not metaphysical, annihilation in view, Buddhism must needs be a sad and gloomy faith seems to be inconsistent with fact; on the contrary, the prospect of Nirvana fills the true believer, not merely with cheerfulness, but with an ecstatic desire to reach it.

Note 10 (p. 68).

The influence of the picture of the personal qualities of Gautama, afforded by the legendary anecdotes which rapidly grew into a biography of the Buddha; and by the birth stories, which coalesced with the current folk-lore, and were intelligible to all the world, doubtless played a great part. Further, although Gautama appears not to have meddled with the caste system, he refused to recognize any distinction, save that of perfection in the way of salvation, among his followers; and by such teaching, no less than by the inculcation of love and benevolence to all sentient beings, he practically levelled every social, political, and racial barrier. A third important condition was the organization of the Buddhists into monastic communities for the stricter professors, while the laity were permitted a wide indulgence in practice and were allowed to hope for accommodation in some of the temporary abodes of bliss. With a few hundred thousand years of immediate paradise in sight, the average man could be content to shut his eyes to what might follow.

Note 11 (p. 69).

In ancient times it was the fashion, even among the Greeks themselves, to derive all Greek wisdom from Eastern sources; not long ago it was as generally denied that Greek philosophy had any connection with Oriental speculation; it seems probable, however, that the truth lies between these extremes.

The Ionian intellectual movement does not stand alone. It is only one of several sporadic indications of the working of some powerful mental ferment over the whole of the area comprised between the Ægean and Northern Hindostan during the eighth, seventh, and sixth centuries before our era. In these three hundred years, prophetism attained its apogee among the 'Semites of Palestine; Zoroasterism grew and became the creed of a conquering race, the Iranic Aryans; Buddhism rose and spread with marvellous rapidity among the Aryans of Hindostan; while scientific naturalism took its rise among the Aryans of Ionia. It would be difficult to find another three centuries which have given birth to four events of equal importance. All the principal existing religions of mankind have grown out of the first three: while the fourth is the little spring, now swollen into the great stream of positive science. So far as physical possibilities go, the prophet Jeremiah and the oldest Ionian philosopher might have met and conversed. If they had done so, they would probably have disagreed a good deal; and it is interesting to reflect that their discussions might have

embraced questions which, at the present day, are
still hotly controverted.

The old Ionian philosophy, then, seems to be only
one of many results of a stirring of the moral and
intellectual life of the Aryan and the Semitic popu-
lations of Western Asia. The conditions of this
general awakening were doubtless manifold; but
there is one which modern research has brought into
great prominence. This is the existence of extremely
ancient and highly advanced societies in the valleys
of the Euphrates and of the Nile.

It is now known that, more than a thousand—
perhaps more than two thousand—years before the
sixth century B.C., civilization had attained a relatively high pitch among the Babylonians and the
Egyptians. Not only had painting, sculpture,
architecture, and the industrial arts reached a remarkable development; but in Chaldæa, at any rate,
a vast amount of knowledge had been accumulated
and methodized, in the departments of grammar,
mathematics, astronomy, and natural history. Where
such traces of the scientific spirit are visible,
naturalistic speculation is rarely far off, though, so
far as I know, no remains of an Accadian, or
Egyptian, philosophy, properly so called, have yet
been recovered.

Geographically, Chaldæa occupied a central position among the oldest seats of civilization. Commerce, largely aided by the intervention of those
colossal pedlars, the Phœnicians, had brought Chaldæa
into connection with all of them, for a thousand
years before the epoch at present under consideration.

And in the ninth, eighth, and seventh centuries, the Assyrian, the depositary of Chaldæan civilization, as the Macedonian and the Roman, at a later date, were the depositaries of Greek culture, had added irresistible force to the other agencies for the wide distribution of Chaldæan literature, art, and science.

I confess that I find it difficult to imagine that the Greek immigrants—who stood in somewhat the same relation to the Babylonians and the Egyptians as the later Germanic barbarians to the Romans of the Empire—should not have been immensely influenced by the new life with which they became acquainted. But there is abundant direct evidence of the magnitude of this influence in certain spheres. I suppose it is not doubted that the Greek went to school with the Oriental for his primary instruction in reading, writing, and arithmetic ; and that Semitic theology supplied him with some of his mythological lore. Nor does there now seem to be any question about the large indebtedness of Greek art to that of Chaldæa and that of Egypt.

But the manner of that indebtedness is very instructive. The obligation is clear, but its limits are no less definite. Nothing better exemplifies the indomitable originality of the Greeks than the relations of their art to that of the Orientals. Far from being subdued into mere imitators by the technical excellence of their teachers, they lost no time in bettering the instruction they received, using their models as mere stepping stones on the way to those unsurpassed and unsurpassable achievements which are all their own. The shibboleth of Art is

the human figure. The ancient Chaldæans and
Egyptians, like the modern Japanese, did wonders in
the representation of birds and quadrupeds; they
even attained to something more than respectability
in human portraiture. But their utmost efforts never
brought them within range of the best Greek embodi-
ments of the grace of womanhood, or of the severer
beauty of manhood.

It is worth while to consider the probable effect
upon the acute and critical Greek mind of the conflict
of ideas, social, political, and theological, which arose
out of the conditions of life in the Asiatic colonies.
The Ionian polities had passed through the whole
gamut of social and political changes, from patriarchal
and occasionally oppressive kingship to rowdy and
still more burdensome mobship—no doubt with
infinitely eloquent and copious argumentation, on
both sides, at every stage of their progress towards
that arbitrament of force which settles most political
questions. The marvellous speculative faculty,
latent in the Ionian, had come in contact with
Mesopotamian, Egyptian, Phœnician theologies and
cosmogonies; with the illuminati of Orphism and
the fanatics and dreamers of the Mysteries; possibly
with Buddhism and Zoroasterism; possibly even
with Judaism. And it has been observed that the
mutual contradictions of antagonistic supernatural-
isms are apt to play a large part among the genera-
tive agencies of naturalism.

Thus, various external influences may have contributed to the rise of philosophy among the Ionian
Greeks of the sixth century. But the assimilative

capacity of the Greek mind—its power of Hellenizing whatever it touched—has here worked so effectually, that, so far as I can learn, no indubitable traces of such extraneous contributions are now allowed to exist by the most authoritative historians of Philosophy. Nevertheless, I think it must be admitted that the coincidences between the Heracleito-stoical doctrines and those of the older Hindu philosophy are extremely remarkable. In both, the cosmos pursues an eternal succession of cyclical changes. The great year, answering to the Kalpa, covers an entire cycle from the origin of the universe as a fluid to its dissolution in fire—" Humor initium, ignis exitus mundi," as Seneca has it. In both systems, there is immanent in the cosmos a source of energy, Brahma, or the Logos, which works according to fixed laws. The individual soul is an efflux of this world-spirit, and returns to it. Perfection is attainable only by individual effort, through ascetic discipline, and is rather a state of painlessness than of happiness; if indeed it can be said to be a state of anything, save the negation of perturbing emotion. The hatchment motto " In Cœlo Quies " would serve both Hindu and Stoic ; and absolute quiet is not easily distinguishable from annihilation.

Zoroasterism, which, geographically, occupies a position intermediate between Hellenism and Hinduism, agrees with the latter in recognizing the essential evil of the cosmos ; but differs from both in its intensely anthropomorphic personification of the two antagonistic principles, to the one of which it ascribes all the good ; and, to the other, all the evil.

In fact, it assumes the existence of two worlds, one good and one bad ; the latter created by the evil power for the purpose of damaging the former. The existing cosmos is a mere mixture of the two, and the 'last judgment' is a root-and-branch extirpation of the work of Ahriman.

Note 12 (p. 69).

There is no snare in which the feet of a modern student of ancient lore are more easily entangled, than that which is spread by the similarity of the language of antiquity to modern modes of expression. I do not presume to interpret the obscurest of Greek philosophers ; all I wish is to point out, that his words, in the sense accepted by competent inter-preters, fit modern ideas singularly well.

So far as the general theory of evolution goes there is no difficulty. The aphorism about the river ; the figure of the child playing on the shore ; the kingship and fatherhood of strife, seem decisive. The ὁδός ἄνω κάτω μίη expresses, with singular aptness, the cyclical aspect of the one process of organic evolution in individual plants and animals : yet it may be a question whether the Heracleitean strife included any distinct conception of the struggle for existence. Again, it is tempting to compare the part played by the Heracleitean 'fire' with that ascribed by the moderns to heat, or rather to that cause of motion of which heat is one expression ; and a little ingenuity might find a foreshadowing of the doctrine of the conservation of energy, in the saying that all the

things are changed into fire and fire into all things,
as gold into goods and goods into gold.

Note 13 (p. 71).

Popes lines in the *Essay on Man* (Ep. i. 267–8),

> "All are but parts of one stupendous whole,
> Whose body Nature is, and God the soul,"

simply paraphrase Seneca's "quem in hoc mundo
locum deus obtinet, hunc in homine animus: quod
est illic materia, id nobis corpus est."—(Ep. lxv. 24);
which again is a Latin version of the old Stoical
doctrine, εἰς ἅπαν τοῦ κόσμου μέρος διήκει ὁ νοῦς,
καθάπερ ἀφ' ἡμῶν ἡ ψυχή.

So far as the testimony for the universality of what
ordinary people call 'evil' goes, there is nothing
better than the writings of the Stoics themselves.
They might serve as a storehouse for the epigrams of
the ultra-pessimists. Heracleitus (*circa* 500 B.C.)
says just as hard things about ordinary humanity
as his disciples centuries later; and there really
seems no need to seek for the causes of this dark
view of life in the circumstances of the time of
Alexander's successors or of the early Emperors of
Rome. To the man with an ethical ideal, the world,
including himself, will always seem full of evil.

Note 14 (p. 73).

I use the well-known phrase, but decline responsibility for the libel upon Epicurus, whose doctrines
were far less compatible with existence in a stye

than those of the Cynics. If it were steadily borne
in mind that the conception of the 'flesh' as the
source of evil, and the great saying 'Initium est
salutis notitia peccati,' are the property of Epicurus,
fewer illusions about Epicureanism would pass
muster for accepted truth.

Note 15 (p. 75).

The Stoics said that man was a ζῷον λογικὸν
πολιτικὸν φιλάλληλον, or a rational, a political, and
an altruistic or philanthropic animal. In their
view, his higher nature tended to develop in these
three directions, as a plant tends to grow up into
its typical form. Since, without the introduction of
any consideration of pleasure or pain, whatever
thwarted the realization of its type by the plant
might be said to be bad, and whatever helped it good ;
so virtue, in the Stoical sense, as the conduct which
tended to the attainment of the rational, political,
and philanthropic ideal, was good in itself, and
irrespectively of its emotional concomitants.

Man is an " animal sociale communi bono genitum."
The safety of society depends upon practical recognition of the fact. " Salva autem esse societas nisi
custodia et amore partium non possit," says Seneca.
(*De. Ira,* ii. 31.)

Note 16 (p. 75).

The importance of the physical doctrine of the
Stoics lies in its clear recognition of the universality

of the law of causation, with its corollary, the order of nature : the exact form of that order is an altogether secondary consideration.

Many ingenious persons now appear to consider that the incompatibility of pantheism, of materialism, and of any doubt about the immortality of the soul, with religion and morality, is to be held as an axiomatic truth. I confess that I have a certain difficulty in accepting this dogma. For the Stoics were notoriously materialists and pantheists of the most extreme character ; and while no strict Stoic believed in the eternal duration of the individual soul, some even denied its persistence after death. Yet it is equally certain that of all gentile philosophies, Stoicism exhibits the highest ethical development, is animated by the most religious spirit, and has exerted the profoundest influence upon the moral and religious development not merely of the best men among the Romans, but among the moderns down to our own day.

Seneca was claimed as a Christian and placed among the saints by the fathers of the early Christian Church ; and the genuineness of a correspondence between him and the apostle Paul has been hotly maintained in our own time, by orthodox writers. That the letters, as we possess them, are worthless forgeries is obvious ; and writers as wide apart as Baur and Lightfoot agree that the whole story is devoid of foundation.

The dissertation of the late Bishop of Durham (*Epistle to the Philippians*) is particularly worthy of study, apart from this question, on account of the evidence which it supplies of the numerous similarities of thought between Seneca and the writer of the Pauline epistles. When it is remembered that the writer of the Acts puts a quotation from Aratus, or Cleanthes, into the mouth of the apostle; and that Tarsus was a great seat of philosophical and especially stoical learning (Chrysippus himself was a native of the adjacent town of Sôli), there is no difficulty in understanding the origin of these resemblances. See, on this subject, Sir Alexander Grant's dissertation in his edition of *The Ethics of Aristotle* (where there is an interesting reference to the stoical character of Bishop Butler's ethics), the concluding pages of Dr. Weygoldt's instructive little work *Die Philosophie der Stoa*, and Aubertin's *Sénèque et Saint Paul.*

It is surprising that a writer of Dr. Lightfoot's stamp should speak of Stoicism as a philosophy of 'despair.' Surely, rather, it was a philosophy of men who, having cast off all illusions, and the childishness of despair among them, were minded to endure in patience whatever conditions the cosmic process might create, so long as those conditions were compatible with the progress towards virtue, which alone, for them, conferred a worthy object on existence. There is no note of despair in the stoical declaration that the perfected 'wise man' is the equal of Zeus in everything but the duration of his existence. And, in my judgment, there is as little pride about it, often as it serves for the text of discourses on stoical arrogance. Grant the stoical postulate that there is no good except virtue; grant that the per-

fected wise man is altogether virtuous, in consequence
of being guided in all things by the reason, which is
an effluence of Zeus, and there seems no escape from
the stoical conclusion.

Note 17 (p. 76).

Our "Apathy" carries such a different set of
connotations from its Greek original that I have
ventured on using the latter as a technical term.

Note 18 (p. 77).

Many of the stoical philosophers recommended.
their disciples to take an active share in public
affairs ; and in the Roman world, for several
centuries, the best public men were strongly inclined
to Stoicism.   Nevertheless, the logical tendency of
Stoicism seems to me to be fulfilled only in such men
as Diogenes and Epictetus.

Note 19 (p. 80).

"Criticisms on the Origin of Species," 1864.
Collected Essays, vol. ii. p. 91.   [1894.]

Note 20 (p. 81).

Of course, strictly speaking, social life, and the
ethical process in virtue of which it advances towards
perfection, are part and parcel of the general process
of evolution, just as the gregarious habit of in-

numerable plants and animals, which has been of immense advantage to them, is so. A hive of bees is an organic polity, a society in which the part played by each member is determined by organic necessities. Queens, workers, and drones are, so to speak, castes, divided from one another by marked physical barriers. Among birds and mammals, societies are formed, of which the bond in many cases seems to be purely psychological ; that is to say, it appears to depend upon the liking of the individuals for one another's company. The tendency of individuals to over self-assertion is kept down by fighting. Even in these rudimentary forms of society, love and fear come into play, and enforce a greater or less renunciation of self-will. To this extent the general cosmic process begins to be checked by a rudimentary ethical process, which is, strictly speaking, part of the former, just as the 'governor' in a steam-engine is part of the mechanism of the engine.

Note 21 (p. 82).

See " Government : Anarchy or Regimentation," *Collected Essays*, vol. i. pp. 413—418. It is this form of political philosophy to which I conceive the epithet of 'reasoned savagery' to be strictly applicable. [1894.]

Note 22 (p. 83).

" L'homme n'est qu'un roseau, le plus faible de la nature, mais c'est un roseau pensant. Il ne faut pas que l'univers entier s'arme pour l'écraser. Une vapeur, une goutte d'eau, suffit pour le tuer. Mais quand l'univers l'écraserait, l'homme serait encore plus noble que ce qui le tue, parce qu'il sait qu'il meurt ; et l'avantage que l'univers a sur lui, l'univers n'en sait rien."—*Pensées de Pascal.*

Note 23 (p. 85).

The use of the word " Nature " here may be criticised. Yet the manifestation of the natural tendencies of men is so profoundly modified by training that it is hardly too strong. Consider the suppression of the sexual instinct between near relations.

Note 24 (p. 86).

A great proportion of poetry is addressed by the young to the young ; only the great masters of the art are capable of divining, or think it worth while to enter into, the feelings of retrospective age. The two great poets whom we have so lately lost, Tennyson and Browning, have done this, each in his own inimitable way ; the one in the *Ulysses*, from which I have borrowed ; the other in that wonderful fragment ' Childe Roland to the dark Tower came.'

III

SCIENCE AND MORALS

[1886]

In spite of long and, perhaps, not unjustifiable hesitation, I begin to think that there must be something in telepathy. For evidence, which I may not disregard, is furnished by the last number of the "Fortnightly Review" that among the hitherto undiscovered endowments of the human species, there may be a power even more wonderful than the mystic faculty by which the esoterically Buddhistic sage " upon the farthest mountain in Cathay " reads the inmost thoughts of a dweller within the homely circuit of the London postal district. Great indeed is the insight of such a seer; but how much greater is his who combines the feat of reading, not merely the thoughts of which the thinker is aware, but those of which he knows nothing; who sees him unconsciously drawing the conclusions which he repudiates and

supporting the doctrines which he detests. To
reflect upon the confusion which the working of
such a power as this may introduce into one's
ideas of personality and responsibility is perilous
—madness lies that way. But truth is truth, and
I am almost fain to believe in this magical visibi-
lity of the non-existent when the only alternative
is the supposition that the writer of the article on
" Materialism and Morality " in vol. xl. (1886) of
the " Fortnightly Review," in spite of his manifest
ability and honesty, has pledged himself, so far as
I am concerned, to what, if I may trust my own
knowledge of my own thoughts, must be called a
multitude of errors of the first magnitude.

I so much admire Mr. Lilly's outspokenness, I
am so completely satisfied with the uprightness
of his intentions, that it is repugnant to me to
quarrel with anything he may say ; and I sympa-
thise so warmly with his manly scorn of the
vileness of much that passes under the name of
literature in these times, that I would willingly be
silent under his by no means unkindly exposition
of his theory of my own tenets, if I thought that
such personal abnegation would serve the interest
of the cause we both have at heart. But I cannot
think so. My creed may be an ill-favoured thing,
but it is mine own, as Touchstone says of his lady-
love ; and I have so high an opinion of the solid
virtues of the object of my affections that I cannot
calmly see her personated by a wench who is much

uglier and has no virtue worth speaking of. I hope I should be ready to stand by a falling cause if I had ever adopted it; but suffering for a falling cause, which one has done one's best to bring to the ground, is a kind of martyrdom for which I have no taste. In my opinion, the philosophical theory which Mr. Lilly attributes to me—but which I have over and over again disclaimed—is untenable and destined to extinction; and I not unreasonably demur to being counted among its defenders.

After the manner of a mediæval disputant, Mr. Lilly posts up three theses, which, as he conceives, embody the chief heresies propagated by the late Professor Clifford, Mr. Herbert Spencer, and myself. He says that we agree "(1) in putting aside, as unverifiable, everything which the senses cannot verify; (2) everything beyond the bounds of physical science; (3) everything which cannot be brought into a laboratory and dealt with chemically " (p. 578).

My lamented young friend Clifford, sweetest of natures though keenest of disputants, is out of reach of our little controversies, but his works speak for him, and those who run may read a refutation of Mr. Lilly's assertions in them. Mr. Herbert Spencer, hitherto, has shown no lack either of ability or of inclination to speak for himself; and it would be a superfluity, not to say an impertinence, on my part, to take up the cudgels for him. But, for myself, if my know-

ledge of my own consciousness may be assumed to
be adequate (and I make not the least pretension
to acquaintance with what goes on in my "Un-
bewusstsein"), I may be permitted to observe
that the first proposition appears to me to be not
true; that the second is in the same case; and
that, if there be gradations in untrueness, the
third is so monstrously untrue that it hovers on
the verge of absurdity, even if it does not actually
flounder in that logical limbo.   Thus, to all three
theses, I reply in appropriate fashion, *Nego*—I say
No; and I proceed to state the grounds of that
negation, which the proprieties do not permit me
to make quite so emphatic as I could desire.

Let me begin with the first assertion, that I
" put aside, as unverifiable, everything which the
senses cannot verify."   Can such a statement as
this be seriously made in respect of any human
being?   But I am not appointed apologist for
mankind in general; and confining my observations to myself, I beg leave to point out that, at
this present moment, I entertain an unshakable
conviction that Mr. Lilly is the victim of a patent
and enormous misunderstanding, and that I have
not the slightest intention of putting that conviction aside because I cannot " verify " it either
by touch, or taste, or smell, or hearing, or sight,
which (in the absence of any trace of telepathic
faculty) make up the totality of my senses.

Again, I may venture to admire the clear and

vigorous English in which Mr. Lilly embodies his
views; but the source of that admiration does not
lie in anything which my five senses enable me to
discover in the pages of his article, and of which
an orang-outang might be just as acutely sensible.
No, it lies in an appreciation of literary form and
logical structure by æsthetic and intellectual
faculties which are not senses, and which are not
unfrequently sadly wanting where the senses are
in full vigour. My poor relation may beat me in
the matter of sensation; but I am quite confident
that, when style and syllogisms are to be dealt
with, he is nowhere.

If there is anything in the world which I do
firmly believe in, it is the universal validity of the
law of causation; but that universality cannot be
proved by any amount of experience, let alone
that which comes to us through the senses. And
when an effort of volition changes the current of
my thoughts, or when an idea calls up another
associated idea, I have not the slightest doubt
that the process to which the first of the phenomena, in each case, is due stands in the relation
of cause to the second. Yet the attempt to verify
this belief by sensation would be sheer lunacy.
Now I am quite sure that Mr. Lilly does not
doubt my sanity; and the only alternative seems
to be the admission that his first proposition is
erroneous.

The second thesis charges me with putting

aside "as unverifiable" "everything beyond the bounds of physical science." Again I say, No. Nobody, I imagine, will credit me with a desire to limit the empire of physical science, but I really feel bound to confess that a great many very familiar and, at the same time, extremely important phenomena lie quite beyond its legitimate limits. I cannot conceive, for example, how the phenomena of consciousness, as such and apart from the physical process by which they are called into existence, are to be brought within the bounds of physical science. Take the simplest possible example, the feeling of redness. Physical science tells us that it commonly arises as a consequence of molecular changes propagated from the eye to a certain part of the substance of the brain, when vibrations of the luminiferous ether of a certain character fall upon the retina. Let us suppose the process of physical analysis pushed so far that one could view the last link of this chain of molecules, watch their movements as if they were billiard balls, weigh them, measure them, and know all that is physically knowable about them. Well, even in that case, we should be just as far from being able to include the resulting phenomenon of consciousness, the feeling of redness, within the bounds of physical science, as we are at present. It would remain as unlike the phenomena we know under the names of matter and motion as it is now. If there is any

plain truth upon which I have made it my
business to insist over and over again it is this—
and whether it is a truth or not, my insistence
upon it leaves not a shadow of justification for
Mr. Lilly's assertion.

But I ask in this case also, how is it conceivable
that any man, in possession of all his natural
faculties, should hold such an opinion ? I do not
suppose that I am exceptionally endowed because
I have all my life enjoyed a keen perception of
the beauty offered us by nature and by art. Now
physical science may and probably will, some day,
enable our posterity to set forth the exact physical
concomitants and conditions of the strange rapture
of beauty. But if ever that day arrives, the
rapture will remain, just as it is now, outside and
beyond the physical world ; and, even in the
mental world, something superadded to mere sensation. I do not wish to crow unduly over my
humble cousin the orang, but in the æsthetic
province, as in that of the intellect, I am afraid
he is nowhere. I doubt not he would detect a
fruit amidst a wilderness of leaves where I could
see nothing ; but I am tolerably confident that he
has never been awestruck, as I have been, by the
dim religious gloom, as of a temple devoted to the
earthgods, of the tropical forests which he in-
habits. Yet I doubt not that our poor long-
armed and short-legged friend, as he sits medita-
tively munching his durian fruit, has something

behind that sad Socratic face of his which is utterly "beyond the bounds of physical science." Physical science may know all about his clutching the fruit and munching it and digesting it, and how the physical titillation of his palate is transmitted to some microscopic cells of the gray matter of his brain.   But the feelings of sweetness and of satisfaction which, for a moment, hang out their signal lights in his melancholy eyes, are as utterly outside the bounds of physics as is the "fine frenzy" of a human rhapsodist.

Does Mr. Lilly really believe that, putting me aside, there is any man with the feeling of music in him who disbelieves in the reality of the delight which he derives from it, because that delight lies outside the bounds of physical science, not less than outside the region of the mere sense of hearing?   But, it may be, that he includes music, painting, and sculpture under the head of physical science, and in that case I can only regret I am unable to follow him in his ennoblement of my favourite pursuits.

The third thesis runs that I put aside " as unverifiable" "everything which cannot be brought into a laboratory and dealt with chemically"; and, once more, I say No. This wondrous allegation is no novelty; it has not unfrequently reached me from that region where gentle (or ungentle) dulness so often holds unchecked sway—the pulpit.   But I marvel to find that a

writer of Mr. Lilly's intelligence and good faith
is willing to father such a wastrel. If I am to deal
with the thing seriously, I find myself met by
one of the two horns of a dilemma. Either some
meaning, as unknown to usage as to the diction-
aries, attaches to "laboratory" and "chemical,"
or the proposition is (what am I to say in my sore
need for a gentle and yet appropriate word?)—
well—unhistorical.

Does Mr. Lilly suppose that I put aside "as
unverifiable" all the truths of mathematics, of
philology, of history? And if I do not, will he
have the great goodness to say how the binomial
theorem is to be dealt with "chemically," even
in the best-appointed "laboratory"; or where
the balances and crucibles are kept by which the
various theories of the nature of the Basque
language may be tested; or what reagents
will extract the truth from any given History
of Rome, and leave the errors behind as a
residual calx?

I really cannot answer these questions, and
unless Mr. Lilly can, I think he would do well
hereafter to think more than twice before
attributing such preposterous notions to his
fellow-men, who, after all, as a learned counsel
said, are vertebrated animals.

The whole thing perplexes me much; and
I am sure there must be an explanation which
will leave Mr. Lilly's reputation for common sense

and fair dealing untouched. Can it be—I put
this forward quite tentatively—that Mr. Lilly is
the victim of a confusion, common enough among
thoughtless people, and into which he has fallen
unawares ? Obviously, it is one thing to say
that the logical methods of physical science are of
universal applicability, and quite another to affirm
that all subjects of thought lie within the province of physical science. I have often declared
my conviction that there is only one method by
which intellectual truth can be reached, whether
the subject-matter of investigation belongs to the
world of physics or to the world of consciousness ;
and one of the arguments in favour of the use of
physical science as an instrument of education
which I have oftenest used is that, in my opinion,
it exercises young minds in the appreciation of
inductive evidence better than any other study.
But while I repeat my conviction that the physical
sciences probably furnish the best and most easily
appreciable illustrations of the one and indivisible
mode of ascertaining truth by the use of reason,
I beg leave to add that I have never thought of
suggesting that other branches of knowledge may
not afford the same discipline ; and assuredly I
have never given the slightest ground for the
attribution to me of the ridiculous contention
that there is nothing true outside the bounds of
physical science. Doubtless people who wanted
to say something damaging, without too nice a

regard to its truth or falsehood, have often
enough misrepresented my plain meaning. But
Mr. Lilly is not one of these folks at whom one
looks and passes by, and I can but sorrowfully
wonder at finding him in such company.

So much for the three theses which Mr. Lilly
has nailed on to the page of this Review. I think
I have shown that the first is inaccurate, that the
second is inaccurate, and that the third is inaccurate; and that these three inaccurates constitute one prodigious, though I doubt not unintentional, misrepresentation. If Mr. Lilly and I
were dialectic gladiators, fighting in the arena of
the "Fortnightly," under the eye of an editorial
lanista, for the delectation of the public, my best
tactics would now be to leave the field of battle.
For the question whether I do, or do not, hold
certain opinions is a matter of fact, with regard to
which my evidence is likely to be regarded as
conclusive—at least until such time as the telepathy of the unconscious is more generally recognised.

However, some other assertions are made by
Mr. Lilly which more or less involve matters of
opinion whereof the rights and wrongs are less
easily settled, but in respect of which he seems to
me to err quite as seriously as about the topics
we have been hitherto discussing. And the importance of these subjects leads me to venture upon
saying something about them, even though I am

thereby compelled to leave the safe ground of
personal knowledge.

Before launching the three torpedoes which
have so sadly exploded on board his own ship,
Mr. Lilly says that with whatever "rhetorical
ornaments I may gild my teaching," it is
"Materialism." Let me observe, in passing, that
rhetorical ornament is not in my way, and that
gilding refined gold would, to my mind, be less
objectionable than varnishing the fair face of
truth with that pestilent cosmetic, rhetoric. If I
believed that I had any claim to the title of
"Materialist," as that term is understood in the
language of philosophy and not in that of abuse, I
should not attempt to hide it by any sort of gilding. I have not found reason to care much for
hard names in the course of the last thirty years,
and I am too old to develop a new sensitiveness.
But, to repeat what I have more than once taken
pains to say in the most unadorned of plain
language, I repudiate, as philosophical error, the
doctrine of Materialism as I understand it, just as
I repudiate the doctrine of Spiritualism as Mr.
Lilly presents it, and my reason for thus doing is,
in both cases, the same ; namely, that, whatever
their differences, Materialists and Spiritualists
agree in making very positive assertions about
matters of which I am certain I know nothing,
and about which I believe they are, in truth, just
as ignorant. And further, that, even when their

assertions are confined to topics which lie within
the range of my faculties, they often appear to
me to be in the wrong. And there is yet another
reason for objecting to be identified with either of
these sects; and that is that each is extremely
fond of attributing to the other, by way of reproach, conclusions which are the property of
neither, though they infallibly flow from the
logical development of the first principles of both.
Surely a prudent man is not to be reproached
because he keeps clear of the squabbles of these
philosophical Bianchi and Neri, by refusing to
have anything to do with either?

I understand the main tenet of Materialism to
be that there is nothing in the universe but
matter and force; and that all the phenomena of
nature are explicable by deduction from the properties assignable to these two primitive factors.
That great champion of Materialism whom Mr.
Lilly appears to consider to be an authority in
physical science, Dr. Büchner, embodies this
article of faith on his title-page. *Kraft und Stoff*
—force and matter—are paraded as the Alpha and
Omega of existence. This I apprehend is the
fundamental article of the faith materialistic;
and whosoever does not hold it is condemned by
the more zealous of the persuasion (as I have
some reason to know) to the Inferno appointed
for fools or hypocrites. But all this I heartily
disbelieve; and at the risk of being charged with

wearisome repetition of an old story, I will briefly
give my reasons for persisting in my infidelity.
In the first place, as I have already hinted, it
seems to me pretty plain that there is a third
thing in the universe, to wit, consciousness, which,
in the hardness of my heart or head, I cannot see
to be matter, or force, or any conceivable modification of either, however intimately the manifestations of the phenomena of consciousness may be
connected with the phenomena known as matter
and force. In the second place, the arguments
used by Descartes and Berkeley to show that our
certain knowledge does not extend beyond our
states of consciousness, appear to me to be
as irrefragable now as they did when I first
became acquainted with them some half-century
ago. All the materialistic writers I know of who
have tried to bite that file have simply broken
their teeth. But, if this is true, our one certainty
is the existence of the mental world, and that of
Kraft und Stoff falls into the rank of, at best, a
highly probable hypothesis.

Thirdly, when I was a mere boy, with a per-
verse tendency to think when I ought to have
been playing, my mind was greatly exercised by
this formidable problem, What would become of
things if they lost their qualities ? As the qualities
had no objective existence, and the thing without
qualities was nothing, the solid world seemed
whittled away—to my great horror. As I grew

older, and learned to use the terms matter and
force, the boyish problem was revived, *mutato
nomine*. On the one hand, the notion of matter
without force seemed to resolve the world into a
set of geometrical ghosts, too dead even to jabber.
On the other hand, Boscovich's hypothesis, by
which matter was resolved into centres of force,
was very attractive. But when one tried to think
it out, what in the world became of force considered as an objective entity? Force, even the
most materialistic of philosophers will agree with
the most idealistic, is nothing but a name for the
cause of motion. And if, with Boscovich, I
resolved things into centres of force, then matter
vanished altogether and left immaterial entities
in its place. One might as well frankly accept
Idealism and have done with it.

I must make a confession, even if it be humili-
ating. I have never been able to form the
slightest conception of those " forces " which the
Materialists talk about, as if they had samples of
them many years in bottle. They tell me that
matter consists of atoms, which are separated by
mere space devoid of contents; and that, through
this void, radiate the attractive and repulsive
forces whereby the atoms affect one another. If
anybody can clearly conceive the nature of these
things which not only exist in nothingness, but
pull and push there with great vigour, I envy
him for the possession of an intellect of larger
grasp, not only than mine, but than that of

Leibnitz or of Newton.[1] To me the " chimæra,
bombinans in vacuo quia comedit secundas inten-
tiones" of the schoolmen is a familiar and
domestic creature compared with such " forces."
Besides, by the hypothesis, the forces are not
matter; and thus all that is of any particular consequence in the world turns out to be not matter
on the Materialist's own showing. Let it not be
supposed that I am casting a doubt upon the
propriety of the employment of the terms " atom "
and " force," as they stand among the working
hypotheses of physical science. As formulæ which
can be applied, with perfect precision and great convenience, in the interpretation of nature, their value
is incalculable; but, as real entities, having an objective existence, an indivisible particle which nevertheless occupies space is surely inconceivable; and
with respect to the operation of that atom, where
it is not, by the aid of a " force " resident in
nothingness, I am as little able to imagine it as I
fancy any one else is.

Unless and until anybody will resolve all these
doubts and difficulties for me, I think I have a
right to hold aloof from Materialism. As to
Spiritualism, it lands me in even greater difficul-

---

[1] See the famous *Collection of Papers*, published by Clarke in
1717.   Leibnitz says: " ' Tis also a supernatural thing that
bodies should *attract* one another at a distance without any
intermediate means." And Clarke, on behalf of Newton, caps
this as follows : " That one body should attract another without
any intermediate *means* is, indeed, not a *miracle*, but a contradiction ; for 'tis supposing something to act where it is not."

ties when I want to get change for its notes-of-hand in the solid coin of reality. For the assumed substantial entity, spirit, which is supposed to underlie the phenomena of consciousness, as matter underlies those of physical nature, leaves not even a geometrical ghost when these phenomena are abstracted. And, even if we suppose the existence of such an entity apart from qualities—that is to say, a bare existence—for mind, how does anybody know that it differs from that other entity, apart from qualities, which is the supposed substratum of matter? Spiritualism is, after all, little better than Materialism turned upside down. And if I try to think of the "spirit" which a man, by this hypothesis, carries about under his hat, as something devoid of relation to space, and as something indivisible, even in thought, while it is, at the same time, supposed to be in that place and to be possessed of half a dozen different faculties, I confess I get quite lost.

As I have said elsewhere, if I were forced to choose between Materialism and Idealism, I should elect for the latter; and I certainly would have nothing to do with the effete mythology of Spiritualism. But I am not aware that I am under any compulsion to choose either the one or the other. I have always entertained a strong suspicion that the sage who maintained that man is the measure of the universe was sadly in the wrong; and age and experience have not weakened

that conviction. In following these lines of speculation I am reminded of the quarter-deck walks of
my youth. In taking that form of exercise you
may perambulate through all points of the com-
pass with perfect safety, so long as you keep within
certain limits : forget those limits, in your ardour,
and mere smothering and spluttering, if not worse,
await you. I stick by the deck and throw a life-
buoy now and then to the struggling folk who
have gone overboard; and all I get for my
humanity is the abuse of all whenever they leave
off abusing one another.

Tolerably early in life I discovered that one of
the unpardonable sins, in the eyes of most people,
is for a man to presume to go about unlabelled.
The world regards such a person as the police do
an unmuzzled dog, not under proper control. I
could find no label that would suit me, so, in my
desire to range myself and be respectable, I invented one; and, as the chief thing I was sure of
was that I did not know a great many things that
the —ists and the —ites about me professed to be
familiar with, I called myself an Agnostic. Surely
no denomination could be more modest or more
appropriate; and I cannot imagine why I should
be every now and then haled out of my refuge
and declared sometimes to be a Materialist, sometimes an Atheist, sometimes a Positivist; and
sometimes, alas and alack, a cowardly or reaction-
ary Obscurantist.

I trust that I have, at last, made my case clear,
and that henceforth I shall be allowed to rest in
peace—at least, after a further explanation or two,
which Mr. Lilly proves to me may be necessary.
It has been seen that my excellent critic has
original ideas respecting the meaning of the
words "laboratory" and "chemical"; and, as it
appears to me, his definition of "Materialist" is
quite as much peculiar to himself. For, unless I
misunderstand him, and I have taken pains not to
do so, he puts me down as a Materialist (over and
above the grounds which I have shown to have
no foundation); firstly, because I have said that
consciousness is a function of the brain; and,
secondly, because I hold by determinism. With
respect to the first point, I am not aware that
there is any one who doubts that, in the proper
physiological sense of the word function, consciousness, in certain forms at any rate, is a
cerebral function. In physiology we call function
that effect, or series of effects, which results from
the activity of an organ. Thus, it is the function
of muscle to give rise to motion; and the muscle
gives rise to motion when the nerve which
supplies it is stimulated. If one of the nerve-
bundles in a man's arm is laid bare and a stimulus
is applied to certain of the nervous filaments, the
result will be production of motion in that arm.
If others are stimulated, the result will be the
production of the state of consciousness called

pain. Now, if I trace these last nerve-filaments,
I find them to be ultimately connected with part
of the substance of the brain, just as the others
turn out to be connected with muscular substance. If the production of motion in the one
case is properly said to be the function of the
muscular substance, why is the production of a
state of consciousness in the other case not to be
called a function of the cerebral substance? Once
upon a time, it is true, it was supposed that a
certain "animal spirit" resided in muscle and was
the real active agent. But we have done with
that wholly superfluous fiction so far as the
muscular organs are concerned. Why are we to
retain a corresponding fiction for the nervous
organs?

If it is replied that no physiologist, however
spiritual his leanings, dreams of supposing that
simple sensations require a "spirit" for their
production, then I must point out that we are
all agreed that consciousness is a function of
matter, and that particular tenet must be given
up as a mark of Materialism. Any further argument will turn upon the question, not whether
consciousness is a function of the brain, but
whether all forms of consciousness are so. Again,
I hold it would be quite correct to say that
material changes are the causes of psychical
phenomena (and, as a consequence, that the
organs in which these changes take place have

the production of such phenomena for their
function), even if the spiritualistic hypothesis had
any foundation. For nobody hesitates to say that
an event A is the cause of an event Z, even if
there are as many intermediate terms, known and
unknown, in the chain of causation as there are
letters between A and Z. The man who pulls
the trigger of a loaded pistol placed close to
another's head certainly is the cause of that
other's death, though, in strictness, he " causes "
nothing but the movement of the finger upon the
trigger. And, in like manner, the molecular
change which is brought about in a certain
portion of the cerebral substance by the stimulation of a remote part of the body would be
properly said to be the cause of the consequent
feeling, whatever unknown terms were interposed
between the physical agent and the actual psychical product. Therefore, unless Materialism has
the monopoly of the right use of language, I see
nothing materialistic in the phraseology which I
have employed.

The only remaining justification which Mr. Lilly
offers for dubbing me a Materialist, *malgré moi*,
arises out of a passage which he quotes, in which I
say that the progress of science means the extension of the province of what we call matter and
force, and the concomitant gradual banishment
from all regions of human thought of what we call
spirit and spontaneity. I hold that opinion now,

if anything, more firmly than I did when I gave utterance to it a score of years ago, for it has been justified by subsequent events. But what that opinion has to do with Materialism I fail to discover. In my judgment, it is consistent with the most thorough-going Idealism, and the grounds of that judgment are really very plain and simple.

The growth of science, not merely of physical science, but of all science, means the demonstration of order and natural causation among phenomena which had not previously been brought under those conceptions. Nobody who is acquainted with the progress of scientific thinking in every department of human knowledge, in the course of the last two centuries, will be disposed to deny that immense provinces have been added to the realm of science; or to doubt that the next two centuries will be witnesses of a vastly greater annexation. More particularly in the region of the physiology of the nervous system is it justifiable to conclude from the progress that has been made in analysing the relations between material and psychical phenomena, that vast further advances will be made; and that, sooner or later, all the so-called spontaneous operations of the mind will have, not only their relations to one another, but their relations to physical phenomena, connected in natural series of causes and effects, strictly defined. In other words, while, at present, we know only the nearer moiety of the chain of causes and effects, by which the phenomena we call material give rise to those which we call mental; hereafter, we shall get to the further end of the series.

In my innocence, I have been in the habit of supposing that this is merely a statement of facts, and that the good Bishop Berkeley, if he were alive, would find such facts fit into his system without the least difficulty. That Mr. Lilly should play into the hands of his foes, by declaring that unmistakable facts make for them, is an exemplification of ways that are dark, quite unintelligible to me. Surely Mr. Lilly does not hold that the disbelief in spontaneity—which term, if it has any meaning at all, means uncaused action —is a mark of the beast Materialism? If so, he must be prepared to tackle many of the Cartesians (if not Descartes himself), Spinoza and Leibnitz among the philosophers, Augustine, Thomas Aquinas, Calvin and his followers among theologians, as Materialists—and that surely is a sufficient *reductio ad absurdum* of such a classification.

The truth is, that in his zeal to paint "Materialism," in large letters, on everything he dislikes, Mr. Lilly forgets a very important fact, which, however, must be patent to every one who has paid attention to the history of human thought; and that fact is, that every one of the speculative difficulties which beset Kant's three problems, the existence of a Deity, the freedom of the

will, and immortality, existed ages before anything that can be called physical science, and
would continue to exist if modern physical science
were swept away.   All that physical science has
done has been to make, as it were, visible and
tangible some difficulties that formerly were more
hard of apprehension.   Moreover, these difficulties
exist just as much on the hypothesis of Idealism
as on that of Materialism.

The student of nature, who starts from the
axiom of the universality of the law of causation,
cannot refuse to admit an eternal existence; if he
admits the conservation of energy, he cannot
deny the possibility of an eternal energy; if he
admits the existence of immaterial phenomena
in the form of consciousness, he must admit the
possibility, at any rate, of an eternal series of
such phenomena; and, if his studies have not been
barren of the best fruit of the investigation of
nature, he will have enough sense to see that
when Spinoza says, " Per Deum intelligo ens
absolute infinitum, hoc est substantiam constantem
infinitis attributis," the God so conceived is one
that only a very great fool would deny, even in
his heart.   Physical science is as little Atheistic
as it is Materialistic.

So with respect to immortality.   As physical
science states this problem, it seems to stand thus:
" Is there any means of knowing whether the
series of states of consciousness, which has been

casually associated for threescore years and ten
with the arrangement and movements of innumerable millions of successively different material molecules, can be continued, in like association, with some substance which has not the
properties of matter and force?" As Kant said,
on a like occasion, if anybody can answer that
question, he is just the man I want to see. If he
says that consciousness cannot exist, except in
relation of cause and effect with certain organic
molecules, I must ask how he knows that; and if
he says it can, I must put the same question.
And I am afraid that, like jesting Pilate, I shall
not think it worth while (having but little time
before me) to wait for an answer.

Lastly, with respect to the old riddle of the
freedom of the will. In the only sense in which
the word freedom is intelligible to me—that is to
say, the absence of any restraint upon doing what
one likes within certain limits—physical science
certainly gives no more ground for doubting it
than the common sense of mankind does. And if
physical science, in strengthening our belief in the
universality of causation and abolishing chance as
an absurdity, leads to the conclusions of determinism, it does no more than follow the track of
consistent and logical thinkers in philosophy and
in theology, before it existed or was thought of.
Whoever accepts the universality of the law of
causation as a dogma of philosophy, denies the

existence of uncaused phenomena. And the essence of that which is improperly called the freewill doctrine is that occasionally, at any rate, human volition is self-caused, that is to say, not caused at all; for to cause oneself one must have anteceded oneself—which is, to say the least of it, difficult to imagine.

Whoever accepts the existence of an omniscient Deity as a dogma of theology, affirms that the order of things is fixed from eternity to eternity; for the fore-knowledge of an occurrence means that the occurrence will certainly happen; and the certainty of an event happening is what is meant by its being fixed or fated.[1]

[1] I may cite, in support of this obvious conclusion of sound reasoning, two authorities who will certainly not be regarded lightly by Mr. Lilly. These are Augustine and Thomas Aquinas. The former declares that "Fate" is only an ill-chosen name for Providence.

"Prorsus divina providentia regna constituuntur humana. Quæ si propterea quisquam fato tribuit, quia ipsam Dei voluntatem vel potestatem fati nomine appellat, *sententiam teneat, linguam corrigat*" (Augustinus *De Civitate Dei*, V. c. i.)

The other great doctor of the Catholic Church, "Divus Thomas," as Suarez calls him, whose marvellous grasp and subtlety of intellect seem to me to be almost without a parallel, puts the whole case into a nutshell, when he says that the ground for doing a thing in the mind of the doer is as it were the pre-existence of the thing done :

"Ratio autem alicujus fiendi in mente actoris existens est quædam præ-existentia rei fiendæ in eo" (*Summa*, Qu. xxiii. Art. i.)

If this is not enough, I may further ask what "Materialist" has ever given a better statement of the case for determinism, on theistic grounds, than is to be found in the following passage of the *Summa*, Qu. xiv. Art. xiii.

"Omnia quæ sunt in tempore, sunt Deo ab æterno præsentia, non solum ea ex ratione qua habet rationes rerum apud se

Whoever asserts the existence of an omnipotent Deity, that he made and sustains all things, and is the *causa causarum*, cannot, without a contradiction in terms, assert that there is any cause independent of him; and it is a mere subterfuge to assert that the cause of all things can " permit " one of these things to be an independent cause.

Whoever asserts the combination of omniscience and omnipotence as attributes of the Deity, does implicitly assert predestination. For he who knowingly makes a thing and places it in circumstances the operation of which on that thing he is perfectly acquainted with, does predestine that thing to whatever fate may befall it.

Thus, to come, at last, to the really important part of all this discussion, if the belief in a God is essential to morality, physical science offers no obstacle thereto; if the belief in immortality is essential to morality, physical science has no more to say against the probability of that doctrine than the most ordinary experience has, and it effectually closes the mouths of those who pretend to refute it by objections deduced from merely physical

presentes, ut quidam dicunt, sed quia ejus intuitus fertur ab æterno supra omnia, prout sunt in sua præsentialitate. *Unde manifestum est quod contingentia infallibiliter a Deo cognoscuntur*, in quantum subduntur divino conspectui secundum suam præsentialitatem ; et tamen sunt futura contingentia, suis causis proximis comparata."

[As I have not said that Thomas Aquinas is professedly a determinist, I do not see the bearing of citations from him which may be more or less inconsistent with the foregoing.]

data. Finally, if the belief in the uncausedness
of volition is essential to morality, the student of
physical science has no more to say against that
absurdity than the logical philosopher or theo-
logian. Physical science, I repeat, did not invent
determinism, and the deterministic doctrine would
stand on just as firm a foundation as it does if
there were no physical science. Let any one who
doubts this read Jonathan Edwards, whose de-
monstrations are derived wholly from philosophy
and theology.

Thus, when Mr. Lilly, like another Solomon
Eagle, goes about proclaiming " Woe to this wicked
city," and denouncing physical science as the evil
genius of modern days—mother of materialism,
and fatalism, and all sorts of other condemnable
isms—I venture to beg him to lay the blame on
the right shoulders ; or, at least, to put in the
dock, along with Science, those sinful sisters of
hers, Philosophy and Theology, who, being so
much older, should have known better than the
poor Cinderella of the schools and universities
over which they have so long dominated. No
doubt modern society is diseased enough ; but
then it does not differ from older civilisations in
that respect. Societies of men are fermenting
masses, and, as beer has what the Germans call
" Oberhefe " and " Unterhefe," so every society that
has existed has had its scum at the top and its
dregs at the bottom ; but I doubt if any of the

" ages of faith " had less scum or less dregs, or even
showed a proportionally greater quantity of sound
wholesome stuff in the vat. I think it would
puzzle Mr. Lilly, or any one else, to adduce con-
vincing evidence that, at any period of the world's
history, there was a more widespread sense of
social duty, or a greater sense of justice, or of the
obligation of mutual help, than in this England of
ours. Ah! but, says Mr. Lilly, these are all products of our Christian inheritance ; when Christian
dogmas vanish virtue will disappear too, and the
ancestral ape and tiger will have full play. But
there are a good many people who think it obvious
that Christianity also inherited a good deal from
Paganism and from Judaism; and that, if the
Stoics and the Jews revoked their bequest, the
moral property of Christianity would realise very
little. And, if morality has survived the stripping
off of several sets of clothes which have been
found to fit badly, why should it not be able to
get on very well in the light and handy garments
which Science is ready to provide ?

But this by the way. If the diseases of society
consist in the weakness of its faith in the existence
of the God of the theologians, in a future state,
and in uncaused volitions, the indication, as the
doctors say, is to suppress Theology and Philosophy, whose bickerings about things of which
they know nothing have been the prime cause
and continual sustenance of that evil scepticism

which is the Nemesis of meddling with the unknowable.

Cinderella is modestly conscious of her ignorance of these high matters. She lights the fire,
sweeps the house, and provides the dinner; and
is rewarded by being told that she is a base
creature, devoted to low and material interests.
But in her garret she has fairy visions out of the
ken of the pair of shrews who are quarrelling
down stairs. She sees the order which pervades
the seeming disorder of the world; the great
drama of evolution, with its full share of pity
and terror, but also with abundant goodness and
beauty, unrolls itself before her eyes; and she
learns, in her heart of hearts, the lesson, that the
foundation of morality is to have done, once and
for all, with lying; to give up pretending to
believe that for which there is no evidence, and
repeating unintelligible propositions about things
beyond the possibilities of knowledge.

She knows that the safety of morality lies
neither in the adoption of this or that philosophical speculation, or this or that theological
creed, but in a real and living belief in that fixed
order of nature which sends social disorganisation
upon the track of immorality, as surely as it
sends physical disease after physical trespasses.
And of that firm and lively faith it is her high
mission to be the priestess.

IV

CAPITAL—THE MOTHER OF LABOUR

AN ECONOMICAL PROBLEM DISCUSSED FROM A
PHYSIOLOGICAL POINT OF VIEW

[1890]

THE first act of a new-born child is to draw a
deep breath. In fact, it will never draw a deeper,
inasmuch as the passages and chambers of the
lungs, once distended with air, do not empty
themselves again; it is only a fraction of their
contents which passes in and out with the flow
and the ebb of the respiratory tide. Mechanically,
this act of drawing breath, or inspiration, is of
the same nature as that by which the handles of
a bellows are separated, in order to fill the bellows
with air; and, in like manner, it involves that
expenditure of energy which we call exertion, or
work, or labour. It is, therefore, no mere metaphor
to say that man is destined to a life of toil : the
work of respiration which began with his first
breath ends only with his last ; nor does one born

in the purple get off with a lighter task than the child who first sees light under a hedge.

How is it that the new-born infant is enabled to perform this first instalment of the sentence of life-long labour which no man may escape? Whatever else a child may be, in respect of this particular question, it is a complicated piece of mechanism, built up out of materials supplied by its mother; and in the course of such building-up, provided with a set of motors—the muscles. Each of these muscles contains a stock of substance capable of yielding energy under certain conditions, one of which is a change of state in the nerve fibres connected with it. The powder in a loaded gun is such another stock of substance capable of yielding energy in consequence of a change of state in the mechanism of the lock, which intervenes between the finger of the man who pulls the trigger and the cartridge. If that change is brought about, the potential energy of the powder passes suddenly into actual energy, and does the work of propelling the bullet. The powder, therefore, may be appropriately called *work-stuff*, not only because it is stuff which is easily made to yield work in the physical sense, but because a good deal of work in the economical sense has contributed to its production. Labour was necessary to collect, transport, and purify the raw sulphur and saltpetre; to cut wood and convert it into powdered charcoal; to mix these in-

gredients in the right proportions; to give the
mixture the proper grain, and so on. The powder
once formed part of the stock, or capital, of a
powder-maker: and it is not only certain natural
bodies which are collected and stored in the gun-
powder, but the labour bestowed on the operations
mentioned may be figuratively said to be incor-
porated in it.

In principle, the work-stuff stored in the
muscles of the new-born child is comparable to
that stored in the gun-barrel. The infant is
launched into altogether new surroundings; and
these operate through the mechanism of the
nervous machinery, with the result that the
potential energy of some of the work-stuff in the
muscles which bring about inspiration is suddenly
converted into actual energy; and this, operating
through the mechanism of the respiratory apparatus, gives rise to an act of inspiration. As
the bullet is propelled by the "going off" of
the powder, as it might be said that the ribs are
raised and the midriff depressed by the "going off"
of certain portions of muscular work-stuff. This
work-stuff is part of a stock or capital of that
commodity stored up in the child's organism
before birth, at the expense of the mother; and
the mother has made good her expenditure by
drawing upon the capital of food-stuffs which
furnished her daily maintenance.

Under these circumstances, it does not appear

to me to be open to doubt that the primary act of
outward labour in the series which necessarily
accompany the life of man is dependent upon the
pre-existence of a stock of material which is not
only of use to him, but which is disposed in such
a manner as to be utilisable with facility. And
I further imagine that the propriety of the
application of the term 'capital' to this stock of
useful substance cannot be justly called in
question; inasmuch as it is easy to prove that
the essential constituents of the work-stuff
accumulated in the child's muscles have merely
been transferred from the store of food-stuffs,
which everybody admits to be capital, by means
of the maternal organism to that of the child, in
which they are again deposited to await use.
Every subsequent act of labour, in like manner,
involves an equivalent consumption of the child's
store of work-stuff—its vital capital; and one of
the main objects of the process of breathing is to
get rid of some of the effects of that consumption.
It follows, then, that, even if no other than the
respiratory work were going on in the organism,
the capital of work-stuff, which the child brought
with it into the world, must sooner or later be used
up, and the movements of breathing must come
to an end; just as the see-saw of the piston of a
steam-engine stops when the coal in the fireplace
has burnt away.

Milk, however, is a stock of materials which

essentially consists of savings from the food-stuffs
supplied to the mother. And these savings are
in such a physical and chemical condition that
the organism of the child can easily convert them
into work-stuff. That is to say, by borrowing
directly from the vital capital of the mother,
indirectly from the store in the natural bodies
accessible to her, it can make good the loss of
its own. The operation of borrowing, however,
involves further work; that is, the labour of
sucking, which is a mechanical operation of much
the same nature as breathing. The child thus
pays for the capital it borrows in labour; but as
the value in work-stuff of the milk obtained is
very far greater than the value of that labour,
estimated by the consumption of work-stuff it
involves, the operation yields a large profit to the
infant. The overplus of food-stuff suffices to increase the child's capital of work-stuff; and to
supply not only the materials for the enlargement
of the " buildings and machinery" which is expressed by the child's growth, but also the energy
required to put all these materials together, and
to carry them to their proper places. Thus,
throughout the years of infancy, and so long
thereafter as the youth or man is not thrown
upon his own resources, he lives by consuming
the vital capital provided by others. To use a
terminology which is more common than appropriate, whatever work he performs (and he does

a good deal, if only in mere locomotion) is unproductive.

Let us now suppose the child come to man's
estate in the condition of a wandering savage,
dependent for his food upon what he can pick
up or catch, after the fashion of the Australian
aborigines. It is plain that the place of mother,
as the supplier of vital capital, is now taken by
the fruits, seeds, and roots of plants and by various
kinds of animals. It is they alone which contain
stocks of those substances which can be converted
within the man's organism into work-stuff; and of
the other matters, except air and water, required
to supply the constant consumption of his capital
and to keep his organic machinery going. In no
way does the savage contribute to the production
of these substances. Whatever labour he bestows
upon such vegetable and animal bodies, on the
contrary, is devoted to their destruction ; and it is
a mere matter of accident whether a little labour
yields him a great deal—as in the case, for
example, of a stranded whale; or whether much
labour yields next to nothing—as in times of
long-continued drought. The savage, like the
child, borrows the capital he needs, and, at any
rate, intentionally, does nothing towards repay-
ment ; it would plainly be an improper use of the
word " produce " to say that his labour in hunting
for the roots, or the fruits, or the eggs, or the
grubs and snakes, which he finds and eats, " pro-

duces" or contributes to " produce " them. The
same thing is true of more advanced tribes, who
are still merely hunters, such as the Esquimaux.
They may expend more labour and skill; but it is
spent in destruction.

When we pass from these to men who lead a
purely pastoral life, like the South American
Gauchos, or some Asiatic nomads, there is an
important change. Let us suppose the owner of
a flock of sheep to live on the milk, cheese, and
flesh which they yield. It is obvious that the
flock stands to him in the economic relation of
the mother to the child, inasmuch as it supplies
him with food-stuffs competent to make good the
daily and hourly losses of his capital of work-
stuff. If we imagine our sheep-owner to have
access to extensive pastures and to be troubled
neither by predacious animals nor by rival shepherds, the performance of his pastoral functions
will hardly involve the expenditure of any more
labour than is needful to provide him with the
exercise required to maintain health. And this
is true, even if we take into account the trouble
originally devoted to the domestication of the
sheep. It surely would be a most singular pretension for the shepherd to talk of the flock as
the "produce" of his labour in any but a very
limited sense. In truth, his labour would have
been a mere accessory of production of very little
consequence. Under the circumstances supposed,

a ram and some ewes, left to themselves for a few
years, would probably generate as large a flock;
and the superadded labour of the shepherd would
have little more effect upon their production than
upon that of the blackberries on the bushes about
the pastures. For the most part the increment
would be thoroughly unearned; and, if it is a rule
of absolute political ethics that owners have no
claim upon "betterment" brought about inde-
pendently of their own labour, then the shepherd
would have no claim to at least nine-tenths of
the increase of the flock.

But if the shepherd has no real claim to the
title of "producer," who has? Are the rams and
ewes the true "producers"? Certainly their title
is better if, borrowing from the old terminology of
chemistry, they only claim to be regarded as the
"proximate principles" of production. And yet,
if strict justice is to be dispensed, even they are
to be regarded rather as collectors and distri-
butors than as "producers." For all that they
really do is to collect, slightly modify, and render
easily accessible, the vital capital which already
exists in the green herbs on which they feed, but
in such a form as to be practically out of the
reach of man.

Thus, from an economic point of view, the
sheep are more comparable to confectioners than
to producers. The usefulness of biscuit lies in
the raw flour of which it is made; but raw flour

does not answer as an article of human diet, and biscuit does. So the usefulness of mutton lies mainly in certain chemical compounds which it contains: the sheep gets them out of grass; we cannot live on grass, but we can on mutton.

Now, herbaceous and all other green plants stand alone among terrestrial natural bodies, in so far as, under the influence of light, they possess the power to build up, out of the carbonic acid gas in the atmosphere, water and certain nitrogenous and mineral salts, those substances which in the animal organism are utilised as work-stuff. They are the chief and, for practical purposes, the sole producers of that vital capital which we have seen to be the necessary antecedent of every act of labour. Every green plant is a laboratory in which, so long as the sun shines upon it, materials furnished by the mineral world, gases, water, saline compounds, are worked up into those food-stuffs without which animal life cannot be carried on. And since, up to the present time, synthetic chemistry has not advanced so far as to achieve this feat, the green plant may be said to be the only living worker whose labour directly results in the production of that vital capital which is the necessary antecedent of human labour.[1] Nor is this statement a paradox involving perpetual

[1] It remains to be seen whether the plants which have no chlorophyll, and flourish in darkness, such as the *Fungi*, can live upon purely mineral food.

motion, because the energy by which the plant
does its work is supplied by the sun—the prim-
ordial capitalist so far as we are concerned.   But
it cannot be too strongly impressed upon the
mind that sunshine, air, water, the best soil that
is to be found on the surface of the earth, might
co-exist; yet without plants, there is no known
agency competent to generate the so-called
"protein compounds," by which alone animal life
can be permanently supported.   And not only
are plants thus essential; but, in respect of particular kinds of animals, they must be plants of
a particular nature.   If there were no terrestrial
green plants but, say, cypresses and mosses,
pastoral and agricultural life would be alike
impossible; indeed, it is difficult to imagine the
possibility of the existence of any large animal, as
the labour required to get at a sufficiency of the
store of food-stuffs, contained in such plants as
these, could hardly extract from them an equivalent for the waste involved in that expenditure
of work.

We are compact of dust and air; from that we
set out, and to that complexion must we come
at last.   The plant either directly, or by some
animal intermediary, lends us the capital which
enables us to carry on the business of life, as we
flit through the upper world, from the one term
of our journey to the other.   Popularly, no doubt,
it is permissible to speak of the soil as a " pro-

ducer," just as we may talk of the daily movement of the sun. But, as I have elsewhere remarked, propositions which are to bear any deductive strain that may be put upon them must run the risk of seeming pedantic, rather than that of being inaccurate. And the statement that land, in the sense of cultivable soil, is a producer, or even one of the essentials of economic production, is anything but accurate. The process of water-culture, in which a plant is not "planted" in any soil, but is merely supported in water containing in solution the mineral ingredients essential to that plant, is now thoroughly understood; and, if it were worth while, a crop yielding abundant food-stuffs could be raised on an acre of fresh water, no less than on an acre of dry land. In the Arctic regions, again, land has nothing to do with "production" in the social economy of the Esquimaux, who live on seals and other marine animals; and might, like Proteus, shepherd the flocks of Poseidon if they had a mind for pastoral life. But the seals and the bears are dependent on other inhabitants of the sea, until, somewhere in the series, we come to the minute green plants which float in the ocean, and are the real "producers" by which the whole of its vast animal population is supported.[1]

[1] In some remarkable passages of the *Botany* of Sir James Ross's Antarctic voyage, which took place half a century ago, Sir Joseph Hooker demonstrated the dependence of the animal life of the sea upon the minute, indeed microscopic, plants which float in it : a marvellous example of what may be done

Thus, when we find set forth as an "absolute" truth the statement that the essential factors in economic production are land, capital and labour —when this is offered as an axiom whence all sorts of other important truths may be deduced— it is needful to remember that the assertion is true only with a qualification. Undoubtedly "vital capital" is essential; for, as we have seen, no human work can be done unless it exists, not even that internal work of the body which is necessary to passive life. But, with respect to labour (that is, human labour) I hope to have left no doubt on the reader's mind that, in regard to production, the importance of human labour may be so small as to be almost a vanishing quantity. Moreover, it is certain that there is no approximation to a fixed ratio between the expenditure of labour and the production of that vital capital which is the foundation of all wealth. For, suppose that we introduce into our suppositious pastoral paradise beasts of prey and rival shepherds, the amount of labour thrown upon the sheep-owner may increase almost indefinitely, and its importance as a condition of production may be enormously augmented, while the quantity of produce remains stationary. Compare for a moment the unim-

by water-culture. One might indulge in dreams of cultivating and improving diatoms, until the domesticated bore the same relation to the wild forms, as cauliflowers to the primitive *Brassica oleracea,* without passing beyond the limits of fair scientific speculation.

portance of the shepherd's labour, under the circumstances first defined, with its indispensability in countries in which the water for the sheep has to be drawn from deep wells, or in which the flock has to be defended from wolves or from human depredators. As to land, it has been shown that, except as affording mere room and standing ground, the importance of land, great as it may be, is secondary. The one thing needful for economic production is the green plant, as the sole producer of vital capital from natural inorganic bodies. Men might exist without labour (in the ordinary sense) and without land; without plants they must inevitably perish.

That which is true of the purely pastoral condition is *a fortiori* true of the purely agricultural [1] condition, in which the existence of the cultivator is directly dependent on the production of vital capital by the plants which he cultivates. Here, again, the condition precedent of the work of each year is vital capital. Suppose that a man lives exclusively upon the plants which he cultivates. It is obvious that he must have food-stuffs to live upon, while he prepares the soil for sowing and throughout the period which elapses between this and harvest. These food-stuffs must be yielded by the stock remaining over from former crops.

---

[1] It is a pity that we have no word that signifies plant-culture exclusively. But for the present purpose I may restrict agriculture to that sense.

The result is the same as before—the pre-existence of vital capital is the necessary antecedent of labour. Moreover, the amount of labour which contributes, as an accessory condition, to the production of the crop varies as widely in the case of plant-raising as in that of cattle-raising. With favourable soil, climate and other conditions, it may be very small, with unfavourable, very great, for the same revenue or yield of food-stuffs.

Thus, I do not think it is possible to dispute the following proposition : the existence of any man, or of any number of men, whether organised into a polity or not, depends on the production of food-stuffs (that is, vital capital) readily accessible to man, either directly or indirectly, by plants. But it follows that the number of men who can exist, say for one year, on any given area of land, taken by itself, depends upon the quantity of food-stuffs produced by such plants growing on the area in one year. If a is that quantity, and b the minimum of food-stuffs required for each man, $\frac{a}{b} = n$, the maximum number of men who can exist on the area. Now the amount of production (a) is limited by the extent of area occupied ; by the quantity of sunshine which falls upon the area ; by the range and distribution of temperature ; by the force of the winds ; by the supply of water ; by the composition and the physical characters of the soil ; by animal and vegetable competitors and de-

stroyers. The labour of man neither does, nor
can, produce vital capital; all that it can do is to
modify, favourably or unfavourably, the conditions
of its production. The most important of these—
namely, sunshine, range of daily and nightly
temperature, wind—are practically out of men's
reach.[1] On the other hand, the supply of water,
the physical and chemical qualities of the soil,
and the influences of competitors and destroyers,
can often, though by no means always, be largely
affected by labour and skill. And there is no
harm in calling the effect of such labour " production," if it is clearly understood that " production " in this sense is a very different thing from
the " production " of food-stuffs by a plant.

We have been dealing hitherto with suppositions
the materials of which are furnished by everyday
experience, not with mere *a priori* assumptions.
Our hypothetical solitary shepherd with his flock,
or the solitary farmer with his grain field, are
mere bits of such experience, cut out, as it were,
for easy study. Still borrowing from daily experience, let us suppose that either sheep-owner
or farmer, for any reason that may be imagined,

[1] I do not forget electric lighting, greenhouses and hothouses,
and the various modes of affording shelter against violent winds :
but in regard to production of food-stuffs on the large scale they
may be neglected. Even if synthetic chemistry should effect
the construction of proteids, the Laboratory will hardly enter
into competition with the Farm within any time which the
present generation need trouble itself about.

desires the help of one or more other men ; and
that, in exchange for their labour, he offers so
many sheep, or quarts of milk, or pounds of
cheese, or so many measures of grain, for a year's
service. I fail to discover any *a priori* " rights of
labour " in virtue of which these men may insist
on being employed, if they are not wanted. But,
on the other hand, I think it is clear that there
is only one condition upon which the persons to
whom the offer of these " wages " is made can
accept it ; and that is that the things offered in
exchange for a year's work shall contain at least
as much vital capital as a man uses up in doing
the year's work. For no rational man could
knowingly and willingly accept conditions which
necessarily involve starvation. Therefore there is
an irreducible minimum of wages ; it is such an
amount of vital capital as suffices to replace the
inevitable consumption of the person hired. Now,
surely, it is beyond a doubt that these wages,
whether at or above the irreducible minimum, are
paid out of the capital disposable after the wants
of the owner of the flock or of the crop of grain
are satisfied ; and, from what has been said already,
it follows that there is a limit to the number of
men, whether hired, or brought in in any other
way, who can be maintained by the sheepowner
or landowner out of his own resources. Since no
amount of labour can produce an ounce of food-
stuff beyond the maximum producible by a limited

number of plants, under the most favourable circumstances in regard to those conditions which are not affected by labour, it follows that, if the number of men to be fed increases indefinitely, a time must come when some will have to starve. That is the essence of the so-called Malthusian doctrine; and it is a truth which, to my mind, is as plain as the general proposition that a quantity which constantly increases will, some time or other, exceed any greater quantity the amount of which is fixed.

The foregoing considerations leave no doubt about the fundamental condition of the existence of any polity, or organised society of men, either in a purely pastoral or purely agricultural state, or in any mixture of both states. It must possess a store of vital capital to start with, and the means of repairing the consumption of that capital which takes place as a consequence of the work of the members of the society. And, if the polity occupies a completely isolated area of the earth's surface, the numerical strength of that polity can never exceed the quotient of the maximum quantity of food-stuffs producible by the green plants on that area, in each year, divided by the quantity necessary for the maintenance of each person during the year. But, there is a third mode of existence possible to a polity; it may, conceivably, be neither purely pastoral nor purely agricultural, but purely manufacturing. Let us

suppose three islands, like Gran Canaria, Teneriffe
and Lanzerote, in the Canaries, to be quite cut off
from the rest of the world. Let Gran Canaria
be inhabited by grain-raisers, Teneriffe by cattle-
breeders; while the population of Lanzerote
(which we may suppose to be utterly barren)
consists of carpenters, woollen manufacturers, and
shoemakers. Then the facts of daily experience
teach us that the people of Lanzerote could never
have existed unless they came to the island
provided with a stock of food-stuffs; and that
they could not continue to exist, unless that stock,
as it was consumed, was made up by contributions
from the vital capital of either Gran Canaria, or
Teneriffe, or both. Moreover, the carpenters of
Lanzerote could do nothing, unless they were
provided with wood from the other islands; nor
could the wool spinners and weavers or the
shoemakers work without wool and skins from the
same sources. The wood and the wool and the
skins are, in fact, the capital without which their
work as manufacturers in their respective trades
is impossible—so that the vital and other capital
supplied by Gran Canaria and Teneriffe is most
indubitably the necessary antecedent of the
industrial labour of Lanzerote. It is perfectly
true that by the time the wood, the wool, and the
skins reached Lanzerote a good deal of labour in
cutting. shearing, skinning, transport, and so on,
would have been spent upon them. But this

does not alter the fact that the only " production "
which is essential to the existence of the population of Teneriffe and Gran Canaria is that effected
by the green plants in both islands; and that all
the labour spent upon the raw produce useful in
manufacture, directly or indirectly yielded by
them—by the inhabitants of these islands and
by those of Lanzerote into the bargain—will not
provide one solitary Lanzerotian with a dinner,
unless the Teneriffians and Canariotes happen to
want his goods and to be willing to give some of
their vital capital in exchange for them.

Under the circumstances defined, if Teneriffe
and Gran Canaria disappeared, or if their inhabitants ceased to care for carpentry, clothing, or
shoes, the people of Lanzerote must starve.   But
if they wish to buy, then the Lanzerotians, by
"cultivating" the buyers, indirectly favour the
cultivation of the produce of those buyers.

Thus, if the question is asked whether the
labour employed in manufacture in Lanzerote is
"productive" or "unproductive" there can be only
one reply.   If anybody will exchange vital capital,
or that which can be exchanged for vital capital,
for Lanzerote goods, it is productive; if not, it is
unproductive.

In the case of the manufacturer, the dependence
of labour upon capital is still more intimate than
in that of the herdsman or agriculturist.   When
the latter are once started they can go on, without

troubling themselves about the existence of any
other people. But the manufacturer depends on
pre-existing capital, not only at the beginning, but
at the end of his operations. However great the
expenditure of his labour and of his skill, the
result, for the purpose of maintaining his existence, is just the same as if he had done nothing,
unless there is a customer able and willing to
exchange food-stuffs for that which his labour and
skill have achieved.

There is another point concerning which it is
very necessary to have clear ideas. Suppose a
carpenter in Lanzerote to be engaged in making
chests of drawers. Let us suppose that a, the
timber, and b, the grain and meat needful for the
man's sustenance until he can finish a chest of
drawers, have to be paid for by that chest.
Then the capital with which he starts is represented by $a + b$. He could not start at all unless
he had it; day by day, he must destroy more or
less of the substance and of the general adapta-
bility of a in order to work it up into the special
forms needed to constitute the chest of drawers;
and, day by day, he must use up at least so much
of b as will replace his loss of vital capital by the
work of that day. Suppose it takes the carpenter and his workmen ten days to saw up the
timber, to plane the boards, and to give them the
shape and size proper for the various parts of the
chest of drawers. And suppose that he then

offers his heap of boards to the advancer of $a + b$ as
an equivalent for the wood + ten days' supply of
vital capital ?   The latter will surely say : " No.
I did not ask for a heap of boards. I asked for a
chest of drawers.   Up to this time, so far as I am
concerned, you have done nothing and are as
much in my debt as ever."   And if the carpenter
maintained that he had " virtually " created two-
thirds of a chest of drawers, inasmuch as it would
take only five days more to put together the pieces
of wood, and that the heap of boards ought to
be accepted as the equivalent of two-thirds of his
debt, I am afraid the creditor would regard him
as little better than an impudent swindler.   It
obviously makes no sort of difference whether the
Canariote or Teneriffian buyer advanced the wood
and the food-stuffs, on which the carpenter had to
maintain himself; or whether the carpenter had a
stock of both, the consumption of which must be
recouped by the exchange of a chest of drawers for
a fresh supply.   In the latter case, it is even less
doubtful that, if the carpenter offered his boards
to the man who wanted a chest of drawers, the
latter would laugh in his face.   And if he took
the chest of drawers for himself, then so much of
his vital capital would be sunk in it past recovery.
Again, the payment of goods in a lump, for the
chest of drawers, comes to the same thing as the
payment of daily wages for the fifteen days that

the carpenter was occupied in making it. If, at
the end of each day, the carpenter chose to say to
himself " I have ' virtually ' created, by my day's
labour, a fifteenth of what I shall get for the chest
of drawers—therefore my wages are the produce of
my day's labour "—there is no great harm in such
metaphorical speech, so long as the poor man does
not delude himself into the supposition that it
represents the exact truth.   " Virtually " is apt to
cover more intellectual sins than " charity " does
moral delicts.   After what has been said, it surely
must be plain enough that each day's work has
involved the consumption of the carpenter's vital
capital, and the fashioning of his timber, at the
expense of more or less consumption of those
forms of capital.   Whether the $a + b$ to be exchanged for the chest has been advanced as a *loan*,
or is paid daily or weekly as *wages*, or, at some
later time, as the *price* of a finished commodity—
the essential element of the transaction, and the
only essential element, is, that it must, at least,
effect the replacement of the vital capital consumed.   Neither boards nor chest of drawers are
eatable ; and, so far from the carpenter having
produced the essential part of his wages by each
day's labour, he has merely wasted that labour,
unless somebody who happens to want a chest of
drawers offers to exchange vital capital, or something that can procure it, equivalent to the

amount consumed during the process of manufacture.[1]

That it should be necessary, at this time of day, to set forth such elementary truths as these may well seem strange ; but no one who consults that interesting museum of political delusions, " Progress and Poverty," some of the treasures of which I have already brought to light, will doubt the fact, if he bestows proper attention upon the first book of that widely-read work.   At page 15 it is thus written :

> The proposition I shall endeavour to prove is : that wages, instead of being drawn from capital, are, in reality, drawn from the product of the labour for which they are paid.

Again at page 18 :—

> In every case in which labour is exchanged for commodities, production really precedes enjoyment . . . wages are the earnings—that is to say, the makings—of labour—not the advances of capital.

And the proposition which the author endeavours to disprove is the hitherto generally accepted doctrine

> that labour is maintained and paid out of existing capital, before the product which constitutes the ultimate object is secured (p. 16).

The doctrine respecting the relation of capital and wages, which is thus opposed in " Progress and

---

[1] See the discussion of this subject further on.

Poverty," is that illustrated in the foregoing pages;
the truth of which, I conceive, must be plain to
any one who has apprehended the very simple
arguments by which I have endeavoured to
demonstrate it. One conclusion or the other
must be hopelessly wrong ; and, even at the cost of
going once more over some of the ground traversed
in this essay and that on " Natural and Political
Rights," [1] I propose to show that the error lies with
" Progress and Poverty"; in which work, so far as
political science is concerned, the poverty is, to
my eye, much more apparent than the progress.

To begin at the beginning. The author pro-
pounds a definition of wealth : " Nothing which
nature supplies to man without his labour is
wealth " (p. 28). Wealth consists of " natural substances or products which have been adapted by
human labour to human use or gratification, their
value depending upon the amount of labour which,
upon the average, would be required to produce
things of like kind" (p. 27). The following
examples of wealth are given :—

Buildings, cattle, tools, machinery, agricultural and mineral
products, manufactured goods, ships, waggons, furniture, and
the like (p. 27).

I take it that native metals, coal and brick
clay, are "mineral products"; and I quite believe
that they are properly termed "wealth." But
when a seam of coal crops out at the surface, and

[1] *Collected Essays*, vol. i. pp. 359-382.

lumps of coal are to be had for the picking up; or when native copper lies about in nuggets, or when brick clay forms a superficial stratum, it appears to me that these things are supplied to, nay almost thrust upon, man without his labour. According to the definition, therefore, they are not "wealth." According to the enumeration, however, they are "wealth": a tolerably fair specimen of a contradiction in terms. Or does "Progress and Poverty" really suggest that a coal seam which crops out at the surface is not wealth; but that if somebody breaks off a piece and carries it away, the bestowal of this amount of labour upon that particular lump makes it wealth; while the rest remains "not wealth"? The notion that the value of a thing bears any necessary relation to the amount of labour (average or otherwise) bestowed upon it, is a fallacy which needs no further refutation than it has already received. The average amount of labour bestowed upon warming-pans confers no value upon them in the eyes of a Gold-Coast negro; nor would an Esquimaux give a slice of blubber for the most elaborate of ice-machines.

So much for the doctrine of "Progress and Poverty" touching the nature of wealth. Let us now consider its teachings respecting capital as wealth or a part of wealth. Adam Smith's definition "that part of a man's stock which he expects to yield him a revenue is called his capital" is quoted with approval (p. 32); else-

where capital is said to be that part of wealth
" which is devoted to the aid of production " (p.
28) ; and yet again it is said to be

wealth in course of exchange,[1] understanding exchange to
include, not merely the passing from hand to hand, but also
such transmutations as occur when the reproductive or trans-
forming forces of nature are utilised for the increase of wealth
(p. 32).

But if too much pondering over the possible
senses and scope of these definitions should weary
the reader, he will be relieved by the following
acknowledgment :—

Nor is the definition of capital I have suggested of any
importance (p. 33).

The author informs us, in fact, that he is " not
writing a text-book," thereby intimating his
opinion that it is less important to be clear and
accurate when you are trying to bring about a
political revolution than when a merely academic
interest attaches to the subject treated.   But he
is not busy about anything so serious as a text-
book : no, he " is only attempting to discover the
laws which control a great social problem "—a
mode of expression which indicates perhaps the
high-water mark of intellectual muddlement.   I
have heard, in my time, of " laws " which control
other " laws " ; but this is the first occasion on
which " laws " which " control a problem " have
come under my notice.   Even the disquisitions " of

[1] The italics are the author's.

those flabby writers who have burdened the press
and darkened counsel by numerous volumes which
are dubbed political economy " (p. 28) could hardly
furnish their critics with a finer specimen of that
which a hero of the " Dunciad," by the one flash of
genius recorded of him, called " clotted nonsense."

Doubtless it is a sign of grace that the author
of these definitions should attach no importance
to any of them; but since, unfortunately, his
whole argument turns upon the tacit assumption
that they are important, I may not pass them
over so lightly.   The third I give up.   Why anything should be capital when it is " in course of
exchange," and not be capital under other circumstances, passes my understanding.   We are told
that " that part of a farmer's crop held for sale or
for seed, or to feed his help, in part payment of
wages, would be accounted capital; that held for
the care of his family would not be " (p. 31).   But
I fail to discover any ground of reason or authority
for the doctrine that it is only when a crop is
about to be sold or sown, or given as wages, that
it may be called capital.   On the contrary, whether
we consider custom or reason, so much of it as is
stored away in ricks and barns during harvest,
and remains there to be used in any of these ways
months or years afterwards, is customarily and
rightly termed capital.   Surely, the meaning of the
clumsy phrase that capital is " wealth in the course
of exchange " must be that it is " wealth capable of

being exchanged " against labour or anything else.
That, in fact, is the equivalent of the second
definition, that capital is "that part of wealth
which is devoted to the aid of production."
Obviously, if you possess that for which men will
give labour, you can aid production by means of
that labour. And, again, it agrees with the first
definition (borrowed from Adam Smith) that
capital is "that part of a man's stock which he
expects to yield him a revenue." For a revenue
is both etymologically and in sense a "return."
A man gives his labour in sowing grain, or in
tending cattle, because he expects a "return"—a
"revenue"—in the shape of the increase of the
grain or of the herd ; and also, in the latter case,
in the shape of their labour and manure which
"aid the production" of such increase. The grain
and cattle of which he is possessed immediately
after harvest is his capital ; and his revenue for the
twelvemonth, until the next harvest, is the surplus
of grain and cattle over and above the amount
with which he started. This is disposable for
any purpose for which he may desire to use it,
leaving him just as well off as he was at the
beginning of the year. Whether the man keeps
the surplus grain for sowing more land, and the
surplus cattle for occupying more pasture ; whether
he exchanges them for other commodities, such
as the use of the land (as rent) ; or labour (as
wages) ; or whether he feeds himself and his

family, in no way alters their nature as revenue, or affects the fact that this revenue is merely disposable capital.

That (even apart from etymology) cattle are typical examples of capital cannot be denied ("Progress and Poverty," p. 25); and if we seek for that particular quality of cattle which makes them "capital," neither has the author of "Progress and Poverty" supplied, nor is any one else very likely to supply, a better account of the matter than Adam Smith has done. Cattle are "capital" because they are "stock which yields revenue." That is to say, they afford to their owner a supply of that which he desires to possess. And, in this particular case, the "revenue" is not only desirable, but of supreme importance, inasmuch as it is capable of maintaining human life. The herd yields a revenue of food-stuffs as milk and meat; a revenue of skins; a revenue of manure; a revenue of labour; a revenue of exchangeable commodities in the shape of these things, as well as in that of live cattle. In each and all of these capacities cattle are capital; and, conversely, things which possess any or all of these capacities are capital.

Therefore what we find at page 25 of "Progress and Poverty" must be regarded as a welcome lapse into clearness of apprehension :—

A fertile field, a rich vein of ore, a falling stream which supplies power, may give the possessor advantages equivalent to the

possession of capital ; but to class such things as capital would
be to put an end to the distinction between land and capital.

Just so. But the fatal truth is that these things
are capital; and that there really is no fundamental distinction between land and capital. Is
it denied that a fertile field, a rich vein of ore, or
a falling stream, may form part of a man's stock,
and that, if they do, they are capable of yielding
revenue ? Will not somebody pay a share of the
produce in kind, or in money, for the privilege of
cultivating the first; royalties for that of working
the second ; and a like equivalent for that of
erecting a mill on the third ? In what sense, then,
are these things less " capital " than the buildings
and tools which on page 27 of " Progress and
Poverty " are admitted to be capital ? Is it not
plain that if these things confer " advantages
equivalent to the possession of capital," and if the
" advantage " of capital is nothing but the yielding
of revenue, then the denial that they are capital
is merely a roundabout way of self-contradiction ?

All this confused talk about capital, however,
is lucidity itself compared with the exposition of
the remarkable thesis, " Wages not drawn from
capital, but produced by labour," which occupies
the third chapter of " Progress and Poverty."

If, for instance, I devote my labour to gathering birds' eggs
or picking wild berries, the eggs or berries I thus get are my wages.
Surely no one will contend that, in such a case, wages are
drawn from capital. There is no capital in the case (p. 34).

Nevertheless, those who have followed what has been said in the first part of this essay surely neither will, nor can, have any hesitation about substantially adopting the challenged contention, though they may possibly have qualms as to the propriety of the use of the term " wages." [1] They will have no difficulty in apprehending the fact that birds' eggs and berries are stores of food-stuffs, or vital capital; that the man who devotes his labour to getting them does so at the expense of his personal vital capital; and that, if the eggs and the berries are " wages" for his work, they are so because they enable him to restore to his organism the vital capital which he has consumed in doing the work of collection. So that there is really a great deal of " capital in the case."

Our author proceeds :—

An absolutely naked man, thrown on an island where no human being has before trod, may gather birds' eggs or pick berries (p. 34).

No doubt. But those who have followed my argument thus far will be aware that a man's vital capital does not reside in his clothes; and, therefore, they will probably fail, as completely as I do, to discover the relevancy of the statement.

[1] Not merely on the grounds stated below, but on the strength of Mr. George's own definition. Does the gatherer of eggs, or berries, *produce* them by his labour? If so, what do the hens and the bushes do?

Again :—

> Or, if I take a piece of leather and work it up into a pair of shoes, the shoes are my wages—the reward of my exertion. Surely they are not drawn from capital—either my capital or anybody else's capital—but are brought into existence by the labour of which they became the wages ; and, in obtaining this pair of shoes as the wages of my labour, capital is not even momentarily lessened one iota. For if we call in the idea of capital, my capital at the beginning consists of the piece of leather, the thread, &c. (p. 34).

It takes away one's breath to have such a concatenation of fallacies administered in the space of half a paragraph. It does not seem to have occurred to our economical reformer to imagine whence his "capital at the beginning," the "leather, thread, &c." came. I venture to suppose that leather to have been originally cattle-skin; and since calves and oxen are not flayed alive, the existence of the leather implies the lessening of that form of capital by a very considerable iota. It is, therefore, as sure as anything can be that, in the long run, the shoes are drawn from that which is capital *par excellence ;* to wit, cattle. It is further beyond doubt that the operation of tanning must involve loss of capital in the shape of bark, to say nothing of other losses; and that the use of the awls and knives of the shoemaker involves loss of capital in the shape of the store of iron ; further, the shoemaker has been enabled to do his work not only by the vital capital expended during the time occupied in making the pair of

shoes, but by that expended from the time of his
birth, up to the time that he earned wages that
would keep him alive.

" Progress and Poverty " continues :—

As my labour goes on, value is steadily added until, when my
labour results in the finished shoes, I have my capital plus the
difference in value between the material and the shoes.   In
obtaining this additional value—my wages—how is capital, at
any time, drawn upon ? (p. 34).

In return we may inquire, how can any one
propound such a question ? Capital is drawn
upon all the time. Not only when the shoes are
commenced, but while they are being made, and
until they are either used by the shoemaker himself or are purchased by somebody else ; that is,
exchanged for a portion of another man's capital.
In fact (supposing that the shoemaker does not
want shoes himself), it is the existence of vital
capital in the possession of another person and the
willingness of that person to part with more or
less of it in exchange for the shoes—it is these
two conditions, alone, which prevent the shoemaker from having consumed his capital unpro-
ductively, just as much as if he had spent his
time in chopping up the leather into minute
fragments.

Thus, the examination of the very case selected
by the advocate of the doctrine that labour bestowed upon manufacture, without any intervention of capital, can produce wages, proves to be a

N 2

delusion of the first magnitude; even though it
be supported by the dictum of Adam Smith which
is quoted in its favour (p. 34)—

> The produce of labour constitutes the natural recompense or
> wages of labour.   In that original state of things which precedes
> both the appropriation of land and the accumulation of stock,
> the whole produce of labour belongs to the labourer.   He has
> neither landlord nor master to share with him ("Wealth of
> Nations," ch. viii.).

But the whole of this passage exhibits the influence of the French Physiocrats by whom Adam
Smith was inspired, at their worst; that is to say,
when they most completely forsook the ground of
experience for *a priori* speculation.   The confident
reference to "that original state of things" is
quite in the manner of the *Essai sur l'Inégalité.*
Now, the state of men before the "appropriation
of land" and the "accumulation of stock" must
surely have been that of purely savage hunters.
As, by the supposition, nobody would have
possessed land, certainly no man could have had
a landlord; and, if there was no accumulation of
stock in a transferable form, as surely there could
be no master, in the sense of hirer.   But hirer
and hire (that is, wages) are correlative terms,
like mother and child.   As "child" implies
"mother," so does "hire" or "wages" imply a
"hirer" or "wage-giver."   Therefore, when a man
in "the original state of things" gathered fruit or
killed game for his own sustenance, the fruit or

the game could be called his "wages" only in a
figurative sense; as one sees if the term "hire,"
which has a more limited connotation, is substituted for "wage." If not, it must be assumed
that the savage hired himself to get his own
dinner; whereby we are led to the tolerably
absurd conclusion that, as in the "state of nature"
he was his own employer, the "master" and the
labourer, in that model age, appropriated the produce in equal shares! And if this should be not
enough, it has already been seen that, in the
hunting state, man is not even an accessory of
production of vital capital; he merely consumes
what nature produces.

According to the author of "Progress and
Poverty" political economists have been deluded
by a "fallacy which has entangled some of the
most acute minds in a web of their own spinning."

It is in the use of the term capital in two senses. In the
primary proposition that capital is necessary to the exertion of
productive labour, the term "capital" is understood as including
all food, clothing, shelter, &c. ; whereas in the deductions
finally drawn from it, the term is used in its common and
legitimate meaning of wealth devoted, not to the immediate
gratification of desire, but to the procurement of more wealth—
of wealth in the hands of employers as distinguished from
labourers (p. 40).

I am by no means concerned to defend the
political economists who are thus charged with
blundering; but I shall be surprised to learn that
any have carried the art of self-entanglement to

the degree of perfection exhibited by this passage.
Who has ever imagined that wealth which, in the
hands of an employer, is capital, ceases to be capital
if it is in the hands of a labourer? Suppose
a workman to be paid thirty shillings on Saturday
evening for six days' labour, that thirty shillings
comes out of the employer's capital, and receives
the name of "wages" simply because it is exchanged for labour. In the workman's pocket,
as he goes home, it is a part of his capital, in
exactly the same sense as, half an hour before, it
was part of the employer's capital; he is a
capitalist just as much as if he were a Rothschild.
Suppose him to be a single man, whose cooking
and household matters are attended to by the
people of the house in which he has a room;
then the rent which he pays them out of this
capital is, in part, wages for their labour, and he
is, so far, an employer. If he saves one shilling
out of his thirty, he has, to that extent, added to
his capital when the next Saturday comes round.
And if he puts his saved shillings week by week
into the Savings Bank, the difference between him
and the most bloated of bankers is simply one of
degree.

At page 42, we are confidently told that
"labourers - by receiving wages" cannot lessen
"even temporarily" the "capital of the employer,"
while at page 44 it is admitted that in certain
cases the capitalist "pays out capital in wages."

One would think that the "paying out" of
capital is hardly possible without at least a
" temporary " diminution of the capital from which
payment is made. But " Progress and Poverty "
changes all that by a little verbal legerdemain :—

> For where wages are paid before the object of the labour is
> obtained, or is finished—as in agriculture, where ploughing and
> sowing must precede by several months the harvesting of the
> crop ; as in the erection of buildings, the construction of ships,
> railroads, canals, &c.—it is clear that the owners of the capital
> paid in wages cannot expect an immediate return, but, as the
> phrase is, must "outlay it " or "lie out of it " for a time which
> sometimes amounts to many years. And hence, if first princi-
> ples are not kept in mind, it is easy to jump to the conclusion
> that wages are advanced by capital (p. 44).

Those who have paid attention to the argument
of former parts of this paper may not be able to
understand how, if sound "first principles are
kept in mind," any other conclusion can be
reached, whether by jumping, or by any other
mode of logical progression. But the first principle
which our author " keeps in mind " possesses just
that amount of ambiguity which enables him to
play hocus-pocus with it. It is this; that " the
creation of value does not depend upon the finishing of the product " (p. 44).

There is no doubt that, under certain limitations,
this proposition is correct. It is not true that
"labour always adds to capital by its exertion
before it takes from capital its wages " (p. 44),

but it is true that it may, and often does, produce
that effect.

To take one of the examples given, the construction of a ship. The shaping of the timbers
undoubtedly gives them a value (for a shipbuilder)
which they did not possess before. When they
are put together to constitute the framework of
the ship, there is a still further addition of value
(for a shipbuilder); and when the outside planking
is added, there is another addition (for a shipbuilder). Suppose everything else about the hull
is finished, except the one little item of caulking
the seams, there is no doubt that it has still
more value for a shipbuilder. But for whom else
has it any value, except perhaps for a fire-wood
merchant ? What price will any one who wants
a ship—that is to say, something that will carry a
cargo from one port to another—give for the unfinished vessel which would take water in at
every seam and go down in half an hour, if she
were launched ? Suppose the shipbuilder's capital
to fail before the vessel is caulked, and that he
cannot find another shipbuilder who cares to buy
and finish it, what sort of proportion does the
value created by the labour, for which he has paid
out of his capital, stand to that of his advances ?
Surely no one will give him one-tenth of the
capital disbursed in wages, perhaps not so much
even as the prime cost of the raw materials.
Therefore, though the assertion that " the creation

of value does not depend on the finishing of the product " may be strictly true under certain circumstances, it need not be and is not always true. And, if it is meant to imply or suggest that the creation of value in a manufactured article does not depend upon the finishing of that article, a more serious error could hardly be propounded.

Is there not a prodigious difference in the value of an uncaulked and in that of a finished ship; between the value of a house in which only the tiles of the roof are wanting and a finished house; between that of a clock which only lacks the escapement and a finished clock ?

As ships, house, and clock, the unfinished articles have no value whatever—that is to say, no person who wanted to purchase one of these things, for immediate use, would give a farthing for either. The only value they can have, apart from that of the materials they contain, is that which they possess for some one who can finish them, or for some one who can make use of parts of them for the construction of other things. A man might buy an unfinished house for the sake of the bricks; or he might buy an incomplete clock to use the works for some other piece of machinery.

Thus, though every stage of the labour bestowed on raw material, for the purpose of giving rise to a certain product, confers some additional value on that material in the estimation of those who are

engaged in manufacturing that product, the
ratio of that accumulated value, at any stage of
the process, to the value of the finished product
is extremely inconstant, and often small; while,
to other persons, the value of the unfinished product may be nothing, or even a *minus* quantity.
A house-timber merchant, for example, might
consider that wood which had been worked into
the ribs of a ship was spoiled—that is, had less
value than it had as a log.

According to " Progress and Poverty," there was,
really, no advance of capital while the great St.
Gothard tunnel was cut. Suppose that, as the Swiss
and the Italian halves of the tunnel approached
to within half a kilometre, that half-kilometre had
turned out to be composed of practically impene-
trable rock—would anybody have given a centime
for the unfinished tunnel ?   And if not, how comes
it that " the creation of value does not depend on
the finishing of the product " ?

I think it may be not too much to say that, of
all the political delusions which are current in this
queer world, the very stupidest are those which
assume that labour and capital are necessarily
antagonistic ; that all capital is produced by labour
and therefore, by natural right, is the property of
the labourer; that the possessor of capital is a robber
who preys on the workman and appropriates to himself that which he has had no share in producing.

On the contrary, capital and labour are, necessarily, close allies; capital is never a product of human labour alone; it exists apart from human labour; it is the necessary antecedent of labour; and it furnishes the materials on which labour is employed. The only indispensable form of capital—vital capital—cannot be produced by human labour. All that man can do is to favour its formation by the real producers. There is no intrinsic relation between the amount of labour bestowed on an article and its value in exchange. The claim of labour to the total result of operations which are rendered possible only by capital is simply an *a priori* iniquity.

SOCIAL DISEASES AND WORSE REMEDIES

LETTERS TO THE "TIMES" ON MR. BOOTH'S SCHEME. WITH A PREFACE AND INTRODUCTORY ESSAY

[1891]

PREFACE

THE letters which are here collected together were published in the "Times" in the course of the months of December, 1890, and January, 1891.

The circumstances which led me to write the first letter are sufficiently set forth in its opening sentences; and the materials on which I based my criticisms of Mr. Booth's scheme, in this and in the second letter, were wholly derived from Mr. Booth's book. I had some reason to know, however, that when anybody allows his sense of duty so far to prevail over his sense of the blessedness of peace as to write a letter to the "Times," on

any subject of public interest, his reflections, before he has done with the business, will be very
like those of Johnny Gilpin, "who little thought,
when he set out, of running such a rig." Such
undoubtedly are mine when I contemplate these
twelve documents, and call to mind the distinct addition to the revenue of the Post Office which must
have accrued from the mass of letters and
pamphlets which have been delivered at my
door; to say nothing of the unexpected light
upon my character, motives, and doctrines, which
has been thrown by some of the "Times'" correspondents, and by no end of comments elsewhere.

If self-knowledge is the highest aim of man, I
ought by this time to have little to learn. And
yet, if I am awake, some of my teachers—unable,
perhaps, to control the divine fire of the poetic
imagination which is so closely akin to, if not a
part of, the mythopœic faculty—have surely
dreamed dreams. So far as my humbler and
essentially prosaic faculties of observation and
comparison go, plain facts are against them. But,
as I may be mistaken, I have thought it well to
prefix to the letters (by way of "Prolegomena")
an essay which appeared in the "Nineteenth
Century" for January, 1888, in which the principles that, to my mind, lie at the bottom of the
"social question" are stated. So far as Individualism and Regimental Socialism are concerned, this paper simply emphasizes and expands

the opinions expressed in an address to the members
of the Midland Institute, delivered seventeen years
earlier, and still more fully developed in several
essays published in the " Nineteenth Century " in
1889, which I hope, before long, to republish.[1]
The fundamental proposition which runs
through the writings, which thus extend over a
period of twenty years, is, that the common
a priori doctrines and methods of reasoning about
political and social questions are essentially
vicious; and that argumentation on this basis
leads, with equal logical force, to two contradictory
and extremely mischievous systems, the one that
of Anarchic Individualism, the other that of
despotic or Regimental Socialism. Whether I
am right or wrong, I am at least consistent in
opposing both to the best of my ability. Mr.
Booth's system appears to me, and, as I have
shown, is regarded by Socialists themselves, to be
mere autocratic Socialism, masked by its theological exterior. That the " fantastic " religious
skin will wear away, and the Socialistic reality it
covers will show its real nature, is the expressed
hope of one candid Socialist, and may be fairly
conceived to be the unexpressed belief of the
despotic leader of the new Trades Union, who
has shown his zeal, if not his discretion, in cham-
pioning Mr. Booth's projects. [See Letter VIII.]

[1] See *Collected Essays*, vol. i. p. 290 to end ; and this volume,
p. 147.

Yet another word to commentators upon my letters. There are some who rather chuckle, and some who sneer, at what they seem to consider the dexterity of an "old controversial hand," exhibited by the contrast which I have drawn between the methods of conversion depicted in the New Testament and those pursued by fanatics of the Salvationist type, whether they be such as are now exploited by Mr. Booth, or such as those who, from the time of the Anabaptists, to go no further back, have worked upon similar lines.

Whether such observations were intended to be flattering or sarcastic, I must respectfully decline to accept the compliment, or to apply the sarcasm to myself. I object to obliquity of procedure and ambiguity of speech in all shapes. And I confess that I find it difficult to understand the state of mind which leads any one to suppose, that deep respect for single-minded devotion to high aims is incompatible with the unhesitating conviction that those aims include the propagation of doctrines which are devoid of foundation—perhaps even mischievous.

The most degrading feature of the narrower forms of Christianity (of which that professed by Mr. Booth is a notable example) is their insistence that the noblest virtues, if displayed by those who reject their pitiable formulæ, are, as their pet phrase goes, "splendid sins." But there is, perhaps, one step lower; and that is that men, who profess freedom of thought, should fail to see and appreciate that large soul of goodness which often animates even the fanatical adherents of such tenets. I am sorry for any man who can read the epistles to the Galatians and the Corinthians without yielding a large meed of admiration to the fervent humanity of Paul of Tarsus; who can study the lives of Francis of Assisi, or of Catherine of Siena, without wishing that, for the furtherance of his own ideals, he might be even as they; or who can contemplate unmoved the steadfast veracity and true heroism which loom through the fogs of mystical utterance in George Fox. In all these great men and women there lay the root of the matter; a burning desire to amend the condition of their fellow-men, and to put aside all other things for that end. If, in spite of all the dogmatic helps or hindrances in which they were entangled, these people are not to be held in high honour, who are?

I have never expressed a doubt—for I have none—that, when Mr. Booth left the Methodist connection, and started that organisation of the Salvation Army upon which, comparatively recently, such ambitious schemes of social reform have been grafted, he may have deserved some share of such honour. I do not say that, so far as his personal desires and intentions go, he may not still deserve it.

But the correlate of despotic authority is unlimited responsibility. If Mr. Booth is to take credit for any good that the Army system has effected, he must be prepared to bear blame for its inherent evils. As it seems to me, that has happened to him which sooner or later happens to all despots : he has become the slave of his own creation—the prosperity and glory of the soul-saving machine have become the end, instead of a means, of soul-saving; and to maintain these at the proper pitch, the "General" is led to do things which the Mr. Booth of twenty years ago would probably have scorned.

And those who desire, as I most emphatically desire, to be just to Mr. Booth, however badly they may think of the working of the organisation he has founded, will bear in mind that some astute backers of his probably care little enough for Salvationist religion ; and, perhaps, are not very keen about many of Mr. Booth's projects. I have referred to the rubbing of the hands of the Socialists over Mr. Booth's success ;[1] but, unless I err greatly, there are politicians of a certain school to whom it affords still greater satisfaction. Consider what electioneering agents the captains of the Salvation Army, scattered through all our towns, and directed from a political "bureau" in London, would make! Think how political adversaries could be harassed by our local

[1] See *Letter VIII.*

attorney—" tribune of the people," I mean ; and
how a troublesome man, on the other side, could
be " hunted down " upon any convenient charge,
whether true or false, brought by our Vigilance-
familiar ![1]

I entirely acquit Mr. Booth of any complicity
in far-reaching schemes of this kind ; but I did
not write idly when, in my first letter, I gave no
vague warning of what might grow out of the
organised force, drilled in the habit of unhesitating
obedience, which he has created.

[1] See *Letter II.*

INTRODUCTORY ESSAY

THE STRUGGLE FOR EXISTENCE IN HUMAN SOCIETY

[1888]

THE vast and varied procession of events, which we call Nature, affords a sublime spectacle and an inexhaustible wealth of attractive problems to the speculative observer. If we confine our attention to that aspect which engages the attention of the intellect, nature appears a beautiful and harmonious whole, the incarnation of a faultless logical process, from certain premisses in the past to an inevitable conclusion in the future. But if it be regarded from a less elevated, though more human, point of view; if our moral sympathies are allowed to influence our judgment, and we permit ourselves to criticize our great mother as we criticize one another; then our verdict, at least so far as sentient nature is concerned, can hardly be so favourable.

In sober truth, to those who have made a

study of the phenomena of life as they are
exhibited by the higher forms of the animal
world, the optimistic dogma, that this is the best
of all possible worlds, will seem little better than
a libel upon possibility. It is really only another
instance to be added to the many extant, of the
audacity of *à priori* speculators who, having
created God in their own image, find no difficulty
in assuming that the Almighty must have been
actuated by the same motives as themselves.
They are quite sure that, had any other course
been practicable, He would no more have made
infinite suffering a necessary ingredient of His
handiwork than a respectable philosopher would
have done the like.

But even the modified optimism of the time-
honoured thesis of physico-theology, that the
sentient world is, on the whole, regulated by
principles of benevolence, does but ill stand the
test of impartial confrontation with the facts of
the case. No doubt it is quite true that sentient nature affords hosts of examples of subtle
contrivances directed towards the production of
pleasure or the avoidance of pain ; and it may be
proper to say that these are evidences of benevolence. But if so, why is it not equally proper to
say of the equally numerous arrangements, the no
less necessary result of which is the production
of pain, that they are evidences of malevolence ?

If a vast amount of that which, in a piece of
human workmanship, we should call skill, is

visible in those parts of the organization of a deer to which it owes its ability to escape from beasts of prey, there is at least equal skill displayed in that bodily mechanism of the wolf which enables him to track, and sooner or later to bring down, the deer. Viewed under the dry light of science, deer and wolf are alike admirable; and, if both were non-sentient automata, there would be nothing to qualify our admiration of the action of the one on the other. But the fact that the deer suffers, while the wolf inflicts suffering, engages our moral sympathies. We should call men like the deer innocent and good, men such as the wolf malignant and bad; we should call those who defended the deer and aided him to escape brave and compassionate, and those who helped the wolf in his bloody work base and cruel. Surely, if we transfer these judgments to nature outside the world of man at all, we must do so impartially. In that case, the goodness of the right hand which helps the deer, and the wickedness of the left hand which eggs on the wolf, will neutralize one another: and the course of nature will appear to be neither moral nor immoral, but non-moral.

This conclusion is thrust upon us by analogous facts in every part of the sentient world; yet, inasmuch as it not only jars upon prevalent prejudices, but arouses the natural dislike to that which is painful, much ingenuity has been exercised in devising an escape from it.

From the theological side, we are told that this is a state of probation, and that the seeming injustices and immoralities of nature will be compensated by and by. But how this compensation is to be effected, in the case of the great majority of sentient things, is not clear. I apprehend that no one is seriously prepared to maintain that the ghosts of all the myriads of generations of herbivorous animals which lived during the millions of years of the earth's duration, before the appearance of man, and which have all that time been tormented and devoured by carnivores, are to be compensated by a perennial existence in clover; while the ghosts of carnivores are to go to some kennel where there is neither a pan of water nor a bone with any meat on it. Besides, from the point of view of morality, the last stage of things would be worse than the first. For the carnivores, however brutal and sanguinary, have only done that which, if there is any evidence of contrivance in the world, they were expressly constructed to do. Moreover, carnivores and herbivores alike have been subject to all the miseries incidental to old age, disease, and over-multiplication, and both might well put in a claim for "compensation" on this score.

On the evolutionist side, on the other hand, we are told to take comfort from the reflection that the terrible struggle for existence tends to final good, and that the suffering of the ancestor is paid for by the increased perfection of the progeny. There would be something in this argument if, in

Chinese fashion, the present generation could pay
its debts to its ancestors; otherwise it is not clear
what compensation the *Eohippus* gets for his
sorrows in the fact that, some millions of years
afterwards, one of his descendants wins the Derby.
And, again, it is an error to imagine that evolution
signifies a constant tendency to increased perfection. That process undoubtedly involves a constant
re-modelling of the organism in adaptation to new
conditions; but it depends on the nature of those
conditions whether the direction of the modifi-
cations effected shall be upward or downward.
Retrogressive is as practicable as progressive
metamorphosis. If what the physical philosophers
tell us, that our globe has been in a state of fusion,
and, like the sun, is gradually cooling down, is
true; then the time must come when evolution
will mean adaptation to an universal winter, and
all forms of life will die out, except such low and
simple organisms as the Diatom of the arctic and
antarctic ice and the Protococcus of the red snow.
If our globe is proceeding from a condition in
which it was too hot to support any but the lowest
living thing to a condition in which it will be too
cold to permit of the existence of any others, the
course of life upon its surface must describe a
trajectory like that of a ball fired from a mortar;
and the sinking half of that course is as much a
part of the general process of evolution as the rising.

From the point of view of the moralist the

animal world is on about the same level as a
gladiator's show.   The creatures are fairly well
treated, and set to fight—whereby the strongest,
the swiftest, and the cunningest live to fight
another day.   The spectator has no need to turn
his thumbs down, as no quarter is given.   He
must admit that the skill and training displayed
are wonderful.   But he must shut his eyes if he
would not see that more or less enduring suffering
is the meed of both vanquished and victor.   And
since the great game is going on in every corner
of the world, thousands of times a minute ; since,
were our ears sharp enough, we need not descend
to the gates of hell to hear—

sospiri, pianti, ed alti guai.

. . . . .

Voci alte e fioche, e suon di man con elle

—it seems to follow that, if this world is governed
by benevolence, it must be a different sort of
benevolence from that of John Howard.

But the old Babylonians wisely symbolized
Nature by their great goddess Istar, who combined
the attributes of Aphrodite with those of Ares.
Her terrible aspect is not to be ignored or covered
up with shams ; but it is not the only one.   If
the optimism of Leibnitz is a foolish though
pleasant dream, the pessimism of Schopenhauer is
a nightmare, the more foolish because of its
hideousness.   Error which is not pleasant is
surely the worst form of wrong.

This may not be the best of all possible worlds, but to say that it is the worst is mere petulant nonsense. A worn-out voluptuary may find nothing good under the sun, or a vain and inexperienced youth, who cannot get the moon he cries for, may vent his irritation in pessimistic moanings; but there can be no doubt in the mind of any reasonable person that mankind could, would, and in fact do, get on fairly well with vastly less happiness and far more misery than find their way into the lives of nine people out of ten. If each and all of us had been visited by an attack of neuralgia, or of extreme mental depression, for one hour in every twenty-four—a supposition which many tolerably vigorous people know, to their cost, is not extravagant—the burden of life would have been immensely increased without much practical hindrance to its general course. Men with any manhood in them find life quite worth living under worse conditions than these.

There is another sufficiently obvious fact, which renders the hypothesis that the course of sentient nature is dictated by malevolence quite untenable. A vast multitude of pleasures, and these among the purest and the best, are superfluities, bits of good which are to all appearance unnecessary as inducements to live, and are, so to speak, thrown into the bargain of life. To those who experience them, few delights can be more entrancing than such as are afforded by natural beauty, or by the arts, and especially by music; but they are products of, rather than factors in, evolution, and it is probable that they are known, in any considerable degree, to but a very small proportion of mankind.

The conclusion of the whole matter seems to be that, if Ormuzd has not had his way in this world, neither has Ahriman. Pessimism is as little consonant with the facts of sentient existence as optimism. If we desire to represent the course of nature in terms of human thought, and assume that it was intended to be that which it is, we must say that its governing principle is intellectual and not moral; that it is a materialized logical process, accompanied by pleasures and pains, the incidence of which, in the majority of cases, has not the slightest reference to moral desert. That the rain falls alike upon the just and the unjust, and that those upon whom the Tower of Siloam fell were no worse than their neighbours, seem to be Oriental modes of expressing the same conclusion.

In the strict sense of the word "nature," it denotes the sum of the phenomenal world, of that which has been, and is, and will be; and society, like art, is therefore a part of nature. But it is convenient to distinguish those parts of nature in which man plays the part of immediate cause, as something apart; and, therefore, society, like art, is usefully to be considered as distinct from nature. It is the more desirable, and even necessary, to make this distinction, since society differs from nature in having a definite moral object; whence it comes about that the course shaped by the ethical man—the member of society or citizen—necessarily runs counter to that which the non-ethical man—the primitive savage, or man as a mere member of the animal kingdom—tends to adopt. The latter fights out the struggle for existence to the bitter end, like any other animal; the former devotes his best energies to the object of setting limits to the struggle.[1]

In the cycle of phenomena presented by the life of man, the animal, no more moral end is discernible than in that presented by the lives of the wolf and of the deer. However imperfect the relics of prehistoric men may be, the evidence which they afford clearly tends to the conclusion that, for thousands and thousands of years, before the origin of the oldest known civilizations, men were savages of a very low type. They strove with their enemies and their competitors; they preyed upon things weaker or less cunning than themselves; they were born, multiplied without stint, and died, for thousands of generations, alongside the mammoth, the urus, the lion, and

---

[1] [The reader will observe that this is the argument of the Romanes Lecture, in brief.—1894.]

the hyæna, whose lives were spent in the same way; and they were no more to be praised or blamed, on moral grounds, than their less erect and more hairy compatriots.

As among these, so among primitive men, the weakest and stupidest went to the wall, while the toughest and shrewdest, those who were best fitted to cope with their circumstances, but not the best in any other sense, survived. Life was a continual free fight, and beyond the limited and temporary relations of the family, the Hobbesian war of each against all was the normal state of existence. The human species, like others, plashed and floundered amid the general stream of evolution, keeping its head above water as it best might, and thinking neither of whence nor whither.

The history of civilization—that is, of society— on the other hand, is the record of the attempts which the human race has made to escape from this position. The first men who substituted the state of mutual peace for that of mutual war, whatever the motive which impelled them to take that step, created society. But, in establishing peace, they obviously put a limit upon the struggle for existence. Between the members of that society, at any rate, it was not to be pursued à outrance. And of all the successive shapes which society has taken, that most nearly approaches perfection in which the war of individual against individual is most strictly limited.

The primitive savage, tutored by Istar, appropriated whatever took his fancy, and killed
whomsoever opposed him, if he could. On the
contrary, the ideal of the ethical man is to limit
his freedom of action to a sphere in which he
does not interfere with the freedom of others ; he
seeks the common weal as much as his own ; and,
indeed, as an essential part of his own welfare.
Peace is both end and means with him; and he
founds his life on a more or less complete self-
restraint, which is the negation of the unlimited
struggle for existence. He tries to escape from
his place in the animal kingdom, founded on
the free development of the principle of non-
moral evolution, and to establish a kingdom of
Man, governed upon the principle of moral
evolution. For society not only has a moral
end, but in its perfection, social life, is embodied
morality.

But the effort of ethical man to work towards
a moral end by no means abolished, perhaps has
hardly modified, the deep-seated organic impulses
which impel the natural man to follow his non-
moral course. One of the most essential conditions, if not the chief cause, of the struggle for
existence, is the tendency to multiply without
limit, which man shares with all living things.
It is notable that "increase and multiply" is a
commandment traditionally much older than the
ten ; and that it is, perhaps, the only one which

has been spontaneously and *ex animo* obeyed by
the great majority of the human race. But, in
civilized society, the inevitable result of such
obedience is the re-establishment, in all its intensity, of that struggle for existence—the war of
each against all—the mitigation or abolition of
which was the chief end of social organisation.

It is conceivable that, at some period in the
history of the fabled Atlantis, the production of
food should have been exactly sufficient to meet
the wants of the population, that the makers
of the commodities of the artificer should have
amounted to just the number supportable by
the surplus food of the agriculturists. And, as
there is no harm in adding another monstrous
supposition to the foregoing, let it be imagined
that every man, woman, and child was perfectly
virtuous, and aimed at the good of all as the
highest personal good. In that happy land, the
natural man would have been finally put down
by the ethical man. There would have been
no competition, but the industry of each would
have been serviceable to all ; nobody being vain
and nobody avaricious, there would have been
no rivalries ; the struggle for existence would
have been abolished, and the millennium would
have finally set in. But it is obvious that this
state of things could have been permanent only
with a stationary population. Add ten fresh
mouths ; and as, by the supposition, there was

only exactly enough before, somebody must go
on short rations. The Atlantis society might
have been a heaven upon earth, the whole nation
might have consisted of just men, needing no
repentance, and yet somebody must starve. Reckless Istar, non-moral Nature, would have riven
the ethical fabric. I was once talking with a
very eminent physician [1] about the *vis medicatrix
naturæ*. "Stuff!" said he; "nine times out of
ten nature does not want to cure the man : she
wants to put him in his coffin." And Istar-
Nature appears to have equally little sympathy
with the ends of society. "Stuff! she wants
nothing but a fair field and free play for her
darling the strongest."

Our Atlantis may be an impossible figment,
but the antagonistic tendencies which the fable
adumbrates have existed in every society which
was ever established, and, to all appearance, must
strive for the victory in all that will be. Historians point to the greed and ambition of rulers,
to the reckless turbulence of the ruled, to the
debasing effects of wealth and luxury, and to
the devastating wars which have formed a great
part of the occupation of mankind, as the causes
of the decay of states and the foundering of
old civilisations, and thereby point their story
with a moral. No doubt immoral motives of
all sorts have figured largely among the minor

[1] The late Sir W. Gull.

causes of these events. But beneath all this superficial turmoil lay the deep-seated impulse given by unlimited multiplication. In the swarms of colonies thrown out by Phœnicia and by old Greece; in the *ver sacrum* of the Latin races; in the floods of Gauls and of Teutons which burst over the frontiers of the old civilisation of Europe; in the swaying to and fro of the vast Mongolian hordes in late times, the population problem comes to the front in a very visible shape. Nor is it less plainly manifest in the everlasting agrarian questions of ancient Rome than in the Arreoi societies of the Polynesian Islands.

In the ancient world, and in a large part of that in which we live, the practice of infanticide was, or is, a regular and legal custom; famine, pestilence, and war were and are normal factors in the struggle for existence, and they have served, in a gross and brutal fashion, to mitigate the intensity of the effects of its chief cause.

But, in the more advanced civilisations, the progress of private and public morality has steadily tended to remove all these checks. We declare infanticide murder, and punish it as such; we decree, not quite so successfully, that no one shall die of hunger; we regard death from preventible causes of other kinds as a sort of constructive murder, and eliminate pestilence to the best of our ability; we declaim against the curse of war, and the wickedness of the military spirit, and we are never weary of dilating on the blessedness of peace and the innocent beneficence of Industry. In their moments of expansion, even statesmen and men of business go thus far. The finer spirits look to an ideal *civitas Dei ;* a state when every man, having reached the point of absolute self-negation, and having nothing but moral perfection to strive after, peace will truly reign, not merely among nations, but among men, and the struggle for existence will be at an end.

Whether human nature is competent, under any circumstances, to reach, or even seriously advance towards, this ideal condition, is a question which need not be discussed. It will be admitted that mankind has not yet reached this stage by a very long way, and my business is with the present. And that which I wish to point out is that, so long as the natural man increases and multiplies without restraint, so long will peace and industry not only permit, but they will necessitate, a struggle for existence as sharp as any that ever went on under the *régime* of war. If Istar is to reign on the one hand, she will demand her human sacrifices on the other.

Let us look at home. For seventy years peace and industry have had their way among us with less interruption and under more favourable conditions than in any other country on the face of the earth. The wealth of Crœsus was nothing to that which we have accumulated, and our prosperity has filled the world with envy. But Nemesis did not forget Crœsus : has she forgotten us ?

I think not. There are now 36,000,000 of people in our islands, and every year considerably more than 300,000 are added to our numbers.[1] That is to say, about every hundred seconds, or so, a new claimant to a share in the common stock of maintenance presents him or herself among us. At the present time, the produce of the soil does not suffice to feed half its population. The other moiety has to be supplied with food which must be bought from the people of food-producing countries. That is to say, we have to offer them the things which they want in exchange for the things we want. And the things they want and which we can produce better than they can are mainly manufactures—industrial products.

The insolent reproach of the first Napoleon had a very solid foundation. We not only are, but, under penalty of starvation, we are bound to be, a nation of shopkeepers. But other nations also lie under the same necessity of keeping shop, and some of them deal in the same goods as ourselves.

[1] These numbers are only approximately accurate. In 1881, our population amounted to 35,241,482, exceeding the number in 1871 by 3,396,103. The average annual increase in the decennial period 1871—1881 is therefore 339,610. The number of minutes in a calendar year is 525,600.

Our customers naturally seek to get the most and
the best in exchange for their produce. If our
goods are inferior to those of our competitors, there
is no ground, compatible with the sanity of the
buyers, which can be alleged, why they should not
prefer the latter. And, if that result should ever
take place on a large and general scale, five or six
millions of us would soon have nothing to eat.
We know what the cotton famine was; and we
can therefore form some notion of what a dearth
of customers would be.

Judged by an ethical standard, nothing can be
less satisfactory than the position in which we find
ourselves. In a real, though incomplete, degree
we have attained the condition of peace which is
the main object of social organization; and, for
argument's sake, it may be assumed that we
desire nothing but that which is in itself innocent
and praiseworthy—namely, the enjoyment of the
fruits of honest industry. And lo! in spite of
ourselves, we are in reality engaged in an internecine struggle for existence with our presumably
no less peaceful and well-meaning neighbours.
We seek peace and we do not ensue it. The
moral nature in us asks for no more than is compatible with the general good; the non-moral
nature proclaims and acts upon that fine old
Scottish family motto, " Thou shalt starve ere I
want." Let us be under no illusions, then. So
long as unlimited multiplication goes on, no social
organization which has ever been devised, or is

likely to be devised, no fiddle-faddling with the
distribution of wealth, will deliver society from
the tendency to be destroyed by the reproduction
within itself, in its intensest form, of that struggle
for existence the limitation of which is the object
of society. And however shocking to the moral
sense this eternal competition of man against man
and of nation against nation may be; however
revolting may be the accumulation of misery at
the negative pole of society, in contrast with that
of monstrous wealth at the positive pole;[1] this
state of things must abide, and grow continually
worse, so long as Istar holds her way unchecked.
It is the true riddle of the Sphinx; and every
nation which does not solve it will sooner or later
be devoured by the monster itself has generated.

The practical and pressing question for us, just
now, seems to me to be how to gain time.
" Time brings counsel," as the Teutonic proverb
has it; and wiser folk among our posterity may
see their way out of that which at present looks
like an *impasse*.

It would be folly to entertain any ill-feeling
towards those neighbours and rivals who, like
ourselves, are slaves of Istar; but, if somebody is
to be starved, the modern world has no Oracle of
Delphi to which the nations can appeal for an
indication of the victim. It is open to us to try

---

[1] [It is hard to say whether the increase of the unemployed
poor, or that of the unemployed rich, is the greater social
evil.—1894.]

our fortune; and, if we avoid impending fate, there will be a certain ground for believing that we are the right people to escape. *Securus judicat orbis.*

To this end, it is well to look into the necessary conditions of our salvation by works. They are two, one plain to all the world and hardly needing insistence; the other seemingly not so plain, since too often it has been theoretically and practically left out of sight. The obvious condition is that our produce shall be better than that of others. There is only one reason why our goods should be preferred to those of our rivals—our customers must find them better at the price. That means that we must use more knowledge, skill, and industry in producing them, without a proportionate increase in the cost of production; and, as the price of labour constitutes a large element in that cost, the rate of wages must be restricted within certain limits. It is perfectly true that cheap production and cheap labour are by no means synonymous; but it is also true that wages cannot increase beyond a certain proportion without destroying cheapness. Cheapness, then, with, as part and parcel of cheapness, a moderate price of labour, is essential to our success as competitors in the markets of the world.

The second condition is really quite as plainly indispensable as the first, if one thinks seriously about the matter. It is social stability. Society

is stable, when the wants of its members obtain
as much satisfaction as, life being what it is,
common sense and experience show may be
reasonably expected. Mankind, in general, care
very little for forms of government or ideal
considerations of any sort; and nothing really
stirs the great multitude to break with custom
and incur the manifest perils of revolt except the
belief that misery in this world, or damnation in
the next, or both, are threatened by the continuance of the state of things in which they have
been brought up. But when they do attain that
conviction, society becomes as unstable as a
package of dynamite, and a very small matter
will produce the explosion which sends it back to
the chaos of savagery.

It needs no argument to prove that when the
price of labour sinks below a certain point,
the worker infallibly falls into that condition
which the French emphatically call *la misère*—a
word for which I do not think there is any exact
English equivalent. It is a condition in which
the food, warmth, and clothing which are necessary
for the mere maintenance of the functions of the
body in their normal state cannot be obtained;
in which men, women, and children are forced to
crowd into dens wherein decency is abolished and
the most ordinary conditions of healthful existence are impossible of attainment; in which the
pleasures within reach are reduced to bestiality

and drunkenness; in which the pains accumulate at compound interest, in the shape of starvation, disease, stunted development, and moral degradation; in which the prospect of even steady and honest industry is a life of unsuccessful battling with hunger, rounded by a pauper's grave.

That a certain proportion of the members of every great aggregation of mankind should constantly tend to establish and populate such a Slough of Despond as this is inevitable, so long as some people are by nature idle and vicious, while others are disabled by sickness or accident, or thrown upon the world by the death of their bread-winners. So long as that proportion is restricted within tolerable limits, it can be dealt with; and, so far as it arises only from such causes, its existence may and must be patiently borne. But, when the organization of society, instead of mitigating this tendency, tends to continue and intensify it; when a given social order plainly makes for evil and not for good, men naturally enough begin to think it high time to try a fresh experiment. The animal man, finding that the ethical man has landed him in such a slough, resumes his ancient sovereignty, and preaches anarchy; which is, substantially, a proposal to reduce the social cosmos to chaos, and begin the brute struggle for existence once again.

Any one who is acquainted with the state of the population of all great industrial centres, whether in this or other countries, is aware that, amidst a large and increasing body of that population, *la misère* reigns supreme. I have no pretensions to the character of a philanthropist, and I have a special horror of all sorts of sentimental rhetoric ; I am merely trying to deal with facts, to some extent within my own knowledge, and further evidenced by abundant testimony, as a naturalist; and I take it to be a mere plain truth that, throughout industrial Europe, there is not a single large manufacturing city which is free from a vast mass of people whose condition is exactly that described ; and from a still greater mass who, living just on the edge of the social swamp, are liable to be precipitated into it by any lack of demand for their produce. And, with every addition to the population, the multitude already sunk in the pit and the number of the host sliding towards it continually increase.

Argumentation can hardly be needful to make it clear that no society in which the elements of decomposition are thus swiftly and surely accumulating can hope to win in the race of industries.

Intelligence, knowledge, and skill are undoubtedly conditions of success; but of what avail are they likely to be unless they are backed up by honesty, energy, goodwill, and all the physical and moral faculties that go to the making of manhood, and unless they are stimulated by hope of such reward as men may fairly look to ? And what dweller in the slough of want, dwarfed in body and soul, demoralized, hopeless, can reasonably be expected to possess these qualities ?

Any full and permanent development of the productive powers of an industrial population, then, must be compatible with and, indeed, based upon a social organization which will secure a fair amount of physical and moral welfare to that population; which will make for good and not for evil. Natural science and religious enthusiasm rarely go hand in hand, but on this matter their concord is complete; and the least sympathetic of naturalists can but admire the insight and the devotion of such social reformers as the late Lord Shaftesbury, whose recently published " Life and Letters " gives a vivid picture of the condition of the working classes fifty years ago, and of the pit which our industry, ignoring these plain truths, was then digging under its own feet.

There is, perhaps, no more hopeful sign of progress among us, in the last half-century, than the steadily increasing devotion which has been and is directed to measures for promoting physical and moral welfare among the poorer classes. Sanitary reformers, like most other reformers whom I have had the advantage of knowing, seem to need a good dose of fanaticism, as a sort of moral coca, to keep them up to the mark, and, doubtless, they have made many mistakes; but that the endeavour to improve the condition under

which our industrial population live, to amend the
drainage of densely peopled streets, to provide
baths, washhouses, and gymnasia, to facilitate
habits of thrift, to furnish some provision for
instruction and amusement in public libraries and
the like, is not only desirable from a philanthropic
point of view, but an essential condition of safe
industrial development, appears to me to be indisputable. It is by such means alone, so far as I
can see, that we can hope to check the constant
gravitation of industrial society towards *la misère*,
until the general progress of intelligence and
morality leads men to grapple with the sources of
that tendency. If it is said that the carrying out
of such arrangements as those indicated must
enhance the cost of production, and thus handicap
the producer in the race of competition, I venture,
in the first place, to doubt the fact ; but if it be
so, it results that industrial society has to face
a dilemma, either alternative of which threatens
destruction.

On the one hand, a population the labour of which
is sufficiently remunerated may be physically and
morally healthy and socially stable, but may fail
in industrial competition by reason of the dearness
of its produce. On the other hand, a population
the labour of which is insufficiently remunerated
must become physically and morally unhealthy, and
socially unstable ; and though it may succeed for
a while in industrial competition, by reason of the

cheapness of its produce, it must in the end fall,
through hideous misery and degradation, to utter
ruin.

Well, if these are the only possible alternatives,
let us for ourselves and our children choose the
former, and, if need be, starve like men. But I do
not believe that a stable society made up of
healthy, vigorous, instructed, and self-ruling people
would ever incur serious risk of that fate. They
are not likely to be troubled with many competitors of the same character, just yet; and they may
be safely trusted to find ways of holding their
own.

Assuming that the physical and moral well-
being and the stable social order, which are the
indispensable conditions of permanent industrial
development, are secured, there remains for
consideration the means of attaining that knowledge and skill without which, even then, the
battle of competition cannot be successfully
fought. Let us consider how we stand. A vast
system of elementary education has now been in
operation among us for sixteen years, and has
reached all but a very small fraction of the
population. I do not think that there is any
room for doubt that, on the whole, it has worked
well, and that its indirect no less than its direct
benefits have been immense. But, as might be
expected, it exhibits the defects of all our
educational systems—fashioned as they were to

meet the wants of a bygone condition of society.
There is a widespread and, I think, well-justified
complaint that it has too much to do with books
and too little to do with things. I am as little
disposed as any one can well be to narrow early
education and to make the primary school a mere
annexe of the shop. And it is not so much in
the interests of industry, as in that of breadth of
culture, that I echo the common complaint against
the bookish and theoretical character of our
primary instruction.

If there were no such things as industrial
pursuits, a system of education which does
nothing for the faculties of observation, which
trains neither the eye nor the hand, and is compatible with utter ignorance of the commonest
natural truths, might still be reasonably regarded
as strangely imperfect. And when we consider
that the instruction and training which are
lacking are exactly those which are of most
importance for the great mass of our population,
the fault becomes almost a crime, the more that
there is no practical difficulty in making good
these defects. There really is no reason why
drawing should not be universally taught, and it
is an admirable training for both eye and hand.
Artists are born, not made; but everybody may
be taught to draw elevations, plans, and sections;
and pots and pans are as good, indeed better,
models for this purpose than the Apollo Belvedere.

The plant is not expensive; and there is this excellent quality about drawing of the kind indicated, that it can be tested almost as easily and severely as arithmetic. Such drawings are either right or wrong, and if they are wrong the pupil can be made to see that they are wrong. From the industrial point of view, drawing has the further merit that there is hardly any trade in which the power of drawing is not of daily and hourly utility. In the next place, no good reason, except the want of capable teachers, can be assigned why elementary notions of science should not be an element in general instruction. In this case, again, no expensive or elaborate apparatus is necessary. The commonest thing— a candle, a boy's squirt, a piece of chalk—in the hands of a teacher who knows his business, may be made the starting-points whence children may be led into the regions of science as far as their capacity permits, with efficient exercise of their observational and reasoning faculties on the road. If object lessons often prove trivial failures, it is not the fault of object lessons, but that of the teacher, who has not found out how much the power of teaching a little depends on knowing a great deal, and that thoroughly; and that he has not made that discovery is not the fault of the teachers, but of the detestable system of training them which is widely prevalent.[1]

[1] Training in the use of simple tools is no doubt very desir-

As I have said, I do not regard the proposal to add these to the present subjects of universal instruction as made merely in the interests of industry. Elementary science and drawing are just as needful at Eton (where I am happy to say both are now parts of the regular course) as in the lowest primary school. But their importance in the education of the artisan is enhanced, not merely by the fact that the knowledge and skill thus gained—little as they may amount to—will still be of practical utility to him; but, further, because they constitute an introduction to that special training which is commonly called "technical education."

I conceive that our wants in this last direction may be grouped under three heads: (1) Instruction in the principles of those branches of science and of art which are peculiarly applicable to industrial pursuits, which may be called preliminary scientific education. (2) Instruction in the special branches of such applied science and art, as technical education proper. (3) Instruction of teachers in both these branches. (4) Capacity-catching machinery.

A great deal has already been done in each of these directions, but much remains to be done.

able, on all grounds. From the point of view of "culture," the man whose "fingers are all thumbs" is but a stunted creature. But the practical difficulties in the way of introducing handiwork of this kind into elementary schools appear to me to be considerable.

If elementary education is amended in the way that has been suggested, I think that the school-boards will have quite as much on their hands as they are capable of doing well. The influences under which the members of these bodies are elected do not tend to secure fitness for dealing with scientific or technical education; and it is the less necessary to burden them with an uncongenial task, as there are other organizations, not only much better fitted to do the work, but already actually doing it.

In the matter of preliminary scientific education, the chief of these is the Science and Art Department, which has done more during the last quarter of a century for the teaching of elementary science among the masses of the people than any organization which exists either in this or in any other country. It has become veritably a people's university, so far as physical science is concerned. At the foundation of our old universities they were freely open to the poorest, but the poorest must come to them. In the last quarter of a century, the Science and Art Department, by means of its classes spread all over the country and open to all, has conveyed instruction to the poorest. The University Extension movement shows that our older learned corporations have discovered the propriety of following suit.

Technical education, in the strict sense, has become a necessity for two reasons. The old apprenticeship system has broken down, partly by reason of the changed conditions of industrial life, and partly because trades have ceased to be " crafts," the traditional secrets whereof the master handed down to his apprentices. Invention is constantly changing the face of our industries, so that " use and wont," " rule of thumb," and the like, are gradually losing their importance, while that knowledge of principles which alone can deal successfully with changed conditions is becoming more and more valuable. Socially, the " master " of four or five apprentices is disappearing in favour of the " employer " of forty, or four hundred, or four thousand, " hands," and the odds and ends of technical knowledge, formerly picked up in a shop, are not, and cannot be, supplied in the factory. The instruction formerly given by the master must therefore be more than replaced by the systematic teaching of the technical school.

Institutions of this kind on varying scales of magnitude and completeness, from the splendid edifice set up by the City and Guilds Institute to the smallest local technical school, to say nothing of classes, such as those in technology instituted by the Society of Arts (subsequently taken over by the City Guilds), have been established in various parts of the country, and the movement in favour of their increase and multiplication is rapidly growing in breadth and intensity. But

there is much difference of opinion as to the best
way in which the technical instruction, so generally
desired, should be given. Two courses appear
to be practicable : the one is the establishment of
special technical schools with a systematic and
lengthened course of instruction demanding the
employment of the whole time of the pupils. The
other is the setting afoot of technical classes,
especially evening classes, comprising a short
series of lessons on some special topic, which may
be attended by persons already earning wages in
some branch of trade or commerce.

There is no doubt that technical schools, on
the plan indicated under the first head, are
extremely costly ; and, so far as the teaching of
artizans is concerned, it is very commonly objected
to them that, as the learners do not work under
trade conditions, they are apt to fall into ama-
teurish habits, which prove of more hindrance
than service in the actual business of life. When
such schools are attached to factories under the
direction of an employer who desires to train up
a supply of intelligent workmen, of course this
objection does not apply ; nor can the usefulness
of such schools for the training of future employers and for the higher grade of the employed
be doubtful ; but they are clearly out of the
reach of the great mass of the people, who have
to earn their bread as soon as possible. We must
therefore look to the classes, and especially to

evening classes, as the great instrument for the
technical education of the artizan.   The utility of
such classes has now been placed beyond all
doubt; the only question which remains is to
find the ways and means of extending them.

We are here, as in all other questions of social
organization, met by two diametrically opposed
views.   On the one hand, the methods pursued
in foreign countries are held up as our example.
The state is exhorted to take the matter in hand,
and establish a great system of technical education.   On the other hand, many economists of
the individualist school exhaust the resources of
language in condemning and repudiating, not
merely the interference of the general government
in such matters, but the application of a farthing
of the funds raised by local taxation to these
purposes.   I entertain a strong conviction that,
in this country, at any rate, the State had much
better leave purely technical and trade instruction
alone.   But, although my personal leanings are
decidedly towards the individualists, I have arrived at that conclusion on merely practical
grounds.   In fact, my individualism is rather of a
sentimental sort, and I sometimes think I should
be stronger in the faith if it were less vehemently
advocated.[1]   I am unable to see that civil society

[1] In what follows I am only repeating and emphasizing
opinions which I expressed seventeen years ago, in an Address
to the members of the Midland Institute (republished in *Critiques
and Addresses* in 1873, and in Vol. i. of these *Essays*).   I have

is anything but a corporation established for a moral object—namely, the good of its members—and therefore that it may take such measures as seem fitting for the attainment of that which the general voice decides to be the general good. That the suffrage of the majority is by no means a scientific test of social good and evil is unfortunately too true; but, in practice, it is the only test we can apply, and the refusal to abide by it means anarchy. The purest despotism that ever existed is as much based upon that will of the majority (which is usually submission to the will of a small minority) as the freest republic. Law is the expression of the opinion of the majority; and it is law, and not mere opinion, because the many are strong enough to enforce it.

I am as strongly convinced as the most pronounced individualist can be, that it is desirable that every man should be free to act in every way which does not limit the corresponding freedom of his fellow-man. But I fail to connect that great induction of political science with the practical corollary which is frequently drawn from it: that the State—that is, the people in their corporate capacity—has no business to meddle with anything but the administration of justice and external defence. It appears to me that the

seen no reason to modify them, notwithstanding high authority on the other side.

amount of freedom which incorporate society may
fitly leave to its members is not a fixed quantity,
to be determined *a priori* by deduction from the
fiction called " natural rights "; but that it must
be determined by, and vary with, circumstances.
I conceive it to be demonstrable that the higher
and the more complex the organization of the
social body, the more closely is the life of each
member bound up with that of the whole; and
the larger becomes the category of acts which
cease to be merely self-regarding, and which interfere with the freedom of others more or less
seriously.

If a squatter, living ten miles away from any
neighbour, chooses to burn his house down to get
rid of vermin, there may be no necessity (in the
absence of insurance offices) that the law should
interfere with his freedom of action; his act can
hurt nobody but himself. But, if the dweller in
a street chooses to do the same thing, the State
very properly makes such a proceeding a crime,
and punishes it as such. He does meddle with
his neighbour's freedom, and that seriously. So
it might, perhaps, be a tenable doctrine, that it
would be needless, and even tyrannous, to make
education compulsory in a sparse agricultural
population, living in abundance on the produce of
its own soil; but, in a densely populated manufacturing country, struggling for existence with
competitors, every ignorant person tends to

become a burden upon, and, so far, an infringer of the liberty of, his fellows, and an obstacle to their success. Under such circumstances an education rate is, in fact, a war tax, levied for purposes of defence.

That State action always has been more or less misdirected, and always will be so, is, I believe, perfectly true. But I am not aware that it is more true of the action of men in their corporate capacity than it is of the doings of individuals. The wisest and most dispassionate man in existence, merely wishing to go from one stile in a field to the opposite, will not walk quite straight —he is always going a little wrong, and always correcting himself; and I can only congratulate the individualist who is able to say that his general course of life has been of a less undulatory character. To abolish State action, because its direction is never more than approximately correct, appears to me to be much the same thing as abolishing the man at the wheel altogether, because, do what he will, the ship yaws more or less. "Why should I be robbed of my property to pay for teaching another man's children?" is an individualist question, which is not unfrequently put as if it settled the whole business. Perhaps it does, but I find difficulties in seeing why it should. The parish in which I live makes me pay my share for the paving and lighting of a great many streets that I never pass through;

and I might plead that I am robbed to smooth the way and lighten the darkness of other people. But I am afraid the parochial authorities would not let me off on this plea; and I must confess I do not see why they should.

I cannot speak of my own knowledge, but I have every reason to believe that I came into this world a small reddish person, certainly without a gold spoon in my mouth, and in fact with no discernible abstract or concrete " rights " or property of any description. If a foot was not set upon me, at once, as a squalling nuisance, it was either the natural affection of those about me, which I certainly had done nothing to deserve, or the fear of the law which, ages before my birth, was painfully built up by the society into which I intruded, that prevented that catastrophe. If I was nourished, cared for, taught, saved from the vagabondage of a wastrel, I certainly am not aware that I did anything to deserve those advantages. And, if I possess anything now, it strikes me that, though I may have fairly earned my day's wages for my day's work, and may justly call them my property —yet, without that organization of society, created out of the toil and blood of long generations before my time, I should probably have had nothing but a flint axe and an indifferent hut to call my own ; and even those would be mine only so long as no stronger savage came my way.

So that if society, having, quite gratuitously,

done all these things for me, asks me in turn to
do something towards its preservation—even if
that something is to contribute to the teaching of
other men's children—I really, in spite of all my
individualist leanings, feel rather ashamed to
say no. And if I were not ashamed, I cannot say
that I think that society would be dealing un-
justly with me in converting the moral obligation
into a legal one. There is a manifest unfairness
in letting all the burden be borne by the willing
horse.

It does not appear to me, then, that there is
any valid objection to taxation for purposes of
education; but, in the case of technical schools
and classes, I think it is practically expedient
that such a taxation should be local. Our industrial population accumulates in particular
towns and districts; these districts are those
which immediately profit by technical education;
and it is only in them that we can find the men
practically engaged in industries, among whom
some may reasonably be expected to be competent
judges of that which is wanted, and of the best
means of meeting the want.

In my belief, all methods of technical training
are at present tentative, and, to be successful,
each must be adapted to the special peculiarities
of its locality. This is a case in which we want
twenty years, not of "strong government," but of
cheerful and hopeful blundering; and we may be

thankful if we get things straight in that time.

The principle of the Bill introduced, but dropped, by the Government last session, appears to me to be wise, and some of the objections to it I think are due to a misunderstanding. The Bill proposed in substance to allow localities to tax themselves for purposes of technical education—on the condition that any scheme for such purpose should be submitted to the Science and Art Department, and declared by that department to be in accordance with the intention of the Legislature.

A cry was raised that the Bill proposed to throw technical education into the hands of the Science and Art Department. But, in reality, no power of initiation, nor even of meddling with details, was given to that Department—the sole function of which was to decide whether any plan proposed did or did not come within the limits of " technical education." The necessity for such control, somewhere, is obvious. No legislature, certainly not ours, is likely to grant the power of self-taxation without setting limits to that power in some way ; and it would neither have been practicable to devise a legal definition of technical education, nor commendable to leave the question to the Auditor-General, to be fought out in the law-courts. The only alternative was to leave the decision to an appropriate State authority. If it is asked, what is the need of such control if the people of

the localities are the best judges, the obvious
reply is that there are localities and localities,
and that while Manchester, or Liverpool, or
Birmingham, or Glasgow might, perhaps, be
safely left to do as they thought fit, smaller towns,
in which there is less certainty of full discussion
by competent people of different ways of thinking,
might easily fall a prey to crotcheteers.

Supposing our intermediate science teaching
and our technical schools and classes are established, there is yet a third need to be supplied,
and that is the want of good teachers. And it is
necessary not only to get them, but to keep them
when you have got them.

It is impossible to insist too strongly upon the
fact that efficient teachers of science and of technology are not to be made by the processes in
vogue at ordinary training colleges. The memory
loaded with mere bookwork is not the thing
wanted—is, in fact, rather worse than useless—in
the teacher of scientific subjects. It is absolutely
essential that his mind should be full of knowledge and not of mere learning, and that what he
knows should have been learned in the laboratory
rather than in the library. There are happily
already, both in London and in the provinces,
various places in which such training is to be had,
and the main thing at present is to make it in the
first place accessible, and in the next indispensable,
to those who undertake the business of teaching.

But when the well-trained men are supplied, it must be recollected that the profession of teacher is not a very lucrative or otherwise tempting one, and that it may be advisable to offer special inducements to good men to remain in it. These, however, are questions of detail into which it is unnecessary to enter further.

Last, but not least, comes the question of providing the machinery for enabling those who are by nature specially qualified to undertake the higher branches of industrial work, to reach the position in which they may render that service to the community. If all our educational expenditure did nothing but pick one man of scientific or inventive genius, each year, from amidst the hewers of wood and drawers of water, and give him the chance of making the best of his inborn faculties, it would be a very good investment. If there is one such child among the hundreds of thousands of our annual increase, it would be worth any money to drag him either from the slough of misery, or from the hotbed of wealth, and teach him to devote himself to the service of his people. Here, again, we have made a beginning with our scholarships and the like, and need only follow in the tracks already worn.

The programme of industrial development briefly set forth in the preceding pages is not what Kant calls a "Hirngespinnst," a cobweb spun in the brain of a Utopian philosopher. More

or less of it has taken bodily shape in many parts
of the country, and there are towns of no great
size or wealth in the manufacturing districts
(Keighley, for example) in which almost the whole
of it has, for some time, been carried out, so far as
the means at the disposal of the energetic and
public-spirited men who have taken the matter
in hand permitted. The thing can be done;
I have endeavoured to show good grounds for the
belief that it must be done, and that speedily,
if we wish to hold our own in the war of industry. I doubt not that it will be done, whenever
its absolute necessity becomes as apparent to all
those who are absorbed in the actual business of
industrial life as it is to some of the lookers on.

[Perhaps it is necessary for me to add that
technical education is not here proposed as a
panacea for social diseases, but simply as a
medicament which will help the patient to pass
through an imminent crisis.

An ophthalmic surgeon may recommend an
operation for cataract in a man who is going blind,
without being supposed to undertake that it will
cure him of gout. And I may pursue the
metaphor so far as to remark, that the surgeon
is justified in pointing out that a diet of pork-chops
and burgundy will probably kill his patient,
though he may be quite unable to suggest a mode

of living which will free him from his constitu-
tional disorder.

Mr. Booth asks me, Why do you not propose
some plan of your own ? Really, that is no answer
to my argument that his treatment will make the
patient very much worse. [Note added in *Social
Diseases and Worse Remedies, January,* 1891.]

LETTERS TO THE "TIMES"

ON THE

"DARKEST ENGLAND" SCHEME

I

The " Times," December 1st, 1890

SIR,—A short time ago a generous and philanthropic friend wrote to me, placing at my disposal
a large sum of money for the furtherance of the
vast scheme which the " General " of the Salvation
Army has propounded, if I thought it worthy of
support. The responsibility of advising my benevolent correspondent has weighed heavily upon
me, but I felt that it would be cowardly, as well
as ungracious, to refuse to accept it. I have
therefore studied Mr. Booth's book with some care,
for the purpose of separating the essential from
the accessory features of his project, and I have
based my judgment—I am sorry to say an unfavourable one—upon the *data* thus obtained.
Before communicating my conclusions to my
friend, however, I am desirous to know what
there may be to be said in arrest of that judg-

ment; and the matter is of such vast public
importance that I trust you will aid me by publishing this letter, notwithstanding its length.

There are one or two points upon which I
imagine all thinking men have arrived at the
same convictions as those from which Mr. Booth
starts. It is certain that there is an immense
amount of remediable misery among us; that, in
addition to the poverty, disease, and degradation
which are the consequences of causes beyond
human control, there is a vast, probably a very
much larger, quantity of misery which is the
result of individual ignorance, or misconduct, and
of faulty social arrangements. Further, I think
it is not to be doubted that, unless this remediable
misery is effectually dealt with, the hordes of vice
and pauperism will destroy modern civilization as
effectually as uncivilized tribes of another kind
destroyed the great social organization which
preceded ours. Moreover, I think all will agree
that no reforms and improvements will go to the
root of the evil unless they attack it in its
ultimate source—namely, the motives of the
individual man. Honest, industrious, and self-
restraining men will make a very bad social
organization prosper; while vicious, idle, and
reckless citizens will bring to ruin the best that
ever was, or ever will be, invented.

The leading propositions which are peculiar to
Mr. Booth I take to be these : —

(1) That the only adequate means to such reformation of the individual man is the adoption of that form of somewhat corybantic Christianity of which the soldiers of the Salvation Army are the militant missionaries. This implies the belief that the excitement of the religious emotions (largely by processes described by their employers as "rousing" and "convivial") is a desirable and trustworthy method of permanently amending the conduct of mankind.

I demur to these propositions. I am of opinion that the testimony of history, no less than the cool observation of that which lies within the personal experience of many of us, is wholly adverse to it.

(2) That the appropriate instrument for the propagation and maintenance of this peculiar sacramental enthusiasm is the Salvation Army— a body of devotees, drilled and disciplined as a military organization, and provided with a numerous hierarchy of officers, every one of whom is pledged to blind and unhesitating obedience to the "General," who frankly tells us that the first condition of the service is "implicit, unquestioning obedience." "A telegram from me will send any of them to the uttermost parts of the earth"; every one "has taken service on the express condition that he or she will obey, without questioning, or gainsaying, the orders from headquarters" ("Darkest England," p. 243).

This proposition seems to me to be indisputable. History confirms it. Francis of Assisi and Ignatius Loyola made their great experiments on the same principle. Nothing is more certain than that a body of religious enthusiasts (perhaps we may even say fanatics) pledged to blind obedience to their chief, is one of the most efficient instruments for effecting any purpose that the wit of man has yet succeeded in devising. And I can but admire the insight into human nature which has led Mr. Booth to leave his unquestioning and unhesitating instruments unbound by vows. A volunteer slave is worth ten sworn bondsmen.

(3) That the success of the Salvation Army, with its present force of 9416 officers "wholly engaged in the work," its capital of three quarters of a million, its income of the same amount, its 1375 corps at home, and 1499 in the colonies and foreign countries (Appendix, pp. 3 and 4), is a proof that Divine assistance has been vouchsafed to its efforts.

Here I am not able to agree with the sanguine Commander-in-chief of the new model, whose labours in creating it have probably interfered with his acquisition of information respecting the fate of previous enterprises of like kind.

It does not appear to me that his success is in any degree more remarkable than that of Francis of Assisi or that of Ignatius Loyola, than that of George Fox, or even than that of the Mormons, in our own time. When I observe the discrepancies of the doctrinal foundations from which each of these great movements set out, I find it difficult to suppose that supernatural aid has been given to all of them; still more, that Mr. Booth's smaller measure of success is evidence that it has been granted to him.

But what became of the Franciscan experiment[1]? If there was one rule rather than another on which the founder laid stress, it was that his army of friars should be absolute mendicants, keeping themselves sternly apart from all worldly entanglements. Yet, even before the death of Francis, in 1226, a strong party, headed by Elias of Cortona, the deputy of his own appointment, began to hanker after these very things; and, within thirty years of that time, the Franciscans had become one of the most powerful, wealthy, and worldly corporations in Christendom, with their fingers in every sink of political and social corruption, if so be profit for the order could be fished out of it; their principal interest being to fight their rivals, the Dominicans, and to persecute such of their own brethren as were honest enough to try to carry out their founder's plainest injunctions. We also know what has become of Loyola's experiment. For two centuries the Jesuits have been the hope of the enemies of the Papacy;

[1] See note pp. 245-47.

whenever it becomes too prosperous, they are
sure to bring about a catastrophe by their corrupt
use of the political and social influence which
their organization and their wealth secure.

With these examples of that which may happen
to institutions founded by noble men, with high
aims, in the hands of successors of a different
stamp, armed with despotic authority, before me,
common prudence surely requires that, before advising the handing over of a large sum of money
to the general of a new order of mendicants, I
should ask what guarantee there is that, thirty
years hence, the " General " who then autocrati-
cally controls the action, say, of 100,000 officers
pledged to blind obedience, distributed through
the whole length and breadth of the poorer
classes, and each with his finger on the trigger of
a mine charged with discontent and religious
fanaticism ; with the absolute control, say, of
eight or ten millions sterling of capital and as
many of income ; with barracks in every town,
with estates scattered over the country, and with
settlements in the colonies—will exercise his
enormous powers, not merely honestly, but wisely ?
What shadow of security is there that the person
who wields this uncontrolled authority over many
thousands of men shall use it solely for those
philanthropic and religious objects which, I do not
doubt, are alone in the mind of Mr. Booth ?   Who
is to say that the Salvation Army, in the year

1920, shall not be a replica of what the Franciscan
order had become in the year 1260 ?

The personal character and the intentions of
the founders of such organizations as we are
considering count for very little in the formation
of a forecast of their future ; and if they did, it is
no disrespect to Mr. Booth to say that he is not
the peer of Francis of Assisi. But if Francis's
judgment of men was so imperfect as to permit
him to appoint an ambitious intriguer of the
stamp of Brother Elias his deputy, we have no
right to be sanguine about the perspicacity of Mr.
Booth in a like matter.

Adding to all these considerations the fact that
Mr. Llewelyn Davies, the warmth of whose
philanthropy is beyond question, and in whose
competency and fairness I, for one, place implicit
reliance, flatly denies the boasted success of the
Salvation Army in its professed mission, I have
arrived at the conclusion that, as at present
advised, I cannot be the instrument of carrying
out my friend's proposal.

Mr. Booth has pithily characterised certain
benevolent schemes as doing sixpennyworth of
good and a shilling's worth of harm. I grieve to
say that, in my opinion, the definition exactly fits
his own project. Few social evils are of greater
magnitude than uninstructed and unchastened
religious fanaticism; no personal habit more
surely degrades the conscience and the intellect

than blind and unhesitating obedience to unlimited authority. Undoubtedly, harlotry and
intemperance are sore evils, and starvation is
hard to bear, or even to know of; but the
prostitution of the mind, the soddening of the
conscience, the dwarfing of manhood are worse
calamities. It is a greater evil to have the
intellect of a nation put down by organised
fanaticism; to see its political and industrial
affairs at the mercy of a despot whose chief
thought is to make that fanaticism prevail; to
watch the degradation of men, who should feel
themselves individually responsible for their own
and their country's fates, to mere brute instruments, ready to the hand of a master for any use
to which he may put them.

But that is the end to which, in my opinion,
all such organizations as that to which kindly
people, who do not look to the consequences of
their acts, are now giving their thousands, inevitably tend.   Unless clear proof that I am
wrong is furnished, another thousand shall not
be added by my instrumentality.

I am, Sir, your obedient servant,

T. H. HUXLEY.

NOTE.

An authoritative contemporary historian, Matthew Paris, writes thus of the Minorite, or Franciscan, Friars in England in 1235, just nine years after the death of Francis of Assisi :—

"At this time some of the Minorite brethren, as well as some of the Order of Preachers, unmindful of their profession and the restrictions of their order, impudently entered the territories of some noble monasteries, under pretence of fulfilling their duties of preaching, as if intending to depart after preaching the next day. Under pretence of sickness, or on some other pretext, however, they remained, and, constructing an altar of wood, they placed on it a consecrated stone altar, which they had brought with them, and clandestinely and in a low voice performed mass, and even received the confessions of many of the parishioners, to the prejudice of the priests. . . . And if by chance they were not satisfied with this, they broke forth in insults and threats, reviling every other order except their own, and asserting that all the rest were doomed to damnation, and that they would not spare the soles of their feet till they had exhausted the wealth of their opposers, however great it might be. The religious men, therefore, gave way to them in many points, yielding to avoid scandal, and offending those in power. For they were the councillors and messengers of the nobles, and even secretaries of the Pope, and therefore obtained much

secular favour. Some, however, finding themselves
opposed at the Court of Rome, were restrained by
obvious reasons, and went away in confusion ; for the
Supreme Pontiff, with a scowling look, said to them,
'What means this, my brethren? To what lengths
are you going? Have you not professed voluntary
poverty, and that you would traverse towns and
castles and distant places, as the case required, bare-
footed and unostentatiously in order to preach the
word of God in all humility? And do you now
presume to usurp these estates to yourselves against
the will of the lords of these fees? Your religion
appears to be in a great measure dying away, and
your doctrines to be confuted.' "

Under date of 1243, Matthew writes :—

"For three or four hundred years or more the
monastic order did not hasten to destruction so
quickly as their order [Minorites and Preachers] of
whom now the brothers, twenty-four years having
scarcely elapsed, had first built in England dwellings
which rivalled regal palaces in height. These are
they who daily expose to view their inestimable
treasures, in enlarging their sumptuous edifices, and
erecting lofty walls, thereby impudently transgressing
the limits of their original poverty and violating the
basis of their religion, according to the prophecy of
German Hildegarde. When noblemen and rich men
are at the point of death, whom they know to be
possessed of great riches, they, in their love of gain,
diligently urge them, to the injury and loss of the
ordinary pastors, and extort confessions and hidden
wills, lauding themselves and their own order only,

and placing themselves before all others. So no
faithful man now believes he can be saved, except he
is directed by the counsels of the Preachers and
Minorites." — MATTHEW PARIS's *English History.*
Translated by the Rev. J. A. GILES, 1889, Vol. I.

II

The " Times," December 9th, 1890

SIR,—The purpose of my previous letter about
Mr. Booth's scheme was to arouse the contributors
to the military chest of the Salvation Army to a
clear sense of what they are doing. I thought
it desirable that they should be distinctly aware
that they are setting up and endowing a sect, in
many ways analogous to the " Ranters" and
" Revivalists " of undesirable notoriety in former
times; but with this immensely important difference, that it possesses a strong, far-reaching,
centralized organization, the disposal of the physical, moral, and financial strength of which rests
with an irresponsible chief, who, according to his
own account, is assured of the blind obedience of
nearly 10,000 subordinates. I wish them to ask
themselves, Ought prudent men and good citizens
to aid in the establishment of an organization
which, under sundry, by no means improbable,
contingencies, may easily become a worse and

more dangerous nuisance than the mendicant
friars of the middle ages ? If this is an academic
question, I really do not know what questions
deserve to be called practical. As you divined, I
purposely omitted any consideration of the details
of the Salvationist scheme, and of the principles
which animate those who work it, because I
desired that the public appreciation of the evils,
necessarily inherent in all such plans of despotic
social and religious regimentation should not be
obscured by the raising of points of less comparative, however great absolute, importance.

But it is now time to undertake a more particular criticism of " Darkest England." At the
outset of my examination of that work, I was
startled to find that Mr. Booth had put forward his
scheme with an almost incredibly imperfect knowledge of what had been done and is doing in the
same direction. A simple reader might well imagine
that the author of " Darkest England " posed as
the Columbus, or at any rate the Cortez, of that
region. " Go to Mudie's," he tells us, and you
will be surprised to see how few books there are
upon the social problem. That may or may not
be correct ; but if Mr. Booth had gone to a certain reading-room not far from Mudie's, I undertake to say that the well-informed and obliging
staff of the national library in Bloomsbury would
have provided him with more books on this topic,
in almost all European languages, than he would

read in three months. Has socialism no literature ? And what is socialism but an incarnation
of the social question ? Moreover, I am persuaded that even " Mudie's " resources could have
furnished Mr. Booth with the " Life of Lord
Shaftesbury " and Carlyle's works. Mr. Booth
seems to have undertaken to instruct the world
without having heard of " Past and Present " or of
" Latter-Day Pamphlets " ; though, somewhat late
in the day, a judicious friend calls his attention
to them. To those of my contemporaries on whom,
as on myself, Carlyle's writings on this topic made
an ineffaceable impression forty years ago, who
know that, for all that time, hundreds of able
and devoted men, both clerical and lay, have
worked heart and soul for the permanent amend-
ment of the condition of the poor, Mr. Booth's
"Go to Mudie's" affords an apt measure of the depth
of his preliminary studies. However, I am bound
to admit that these earlier labourers in the field
laboured in such a different fashion, that the originality of the plan started by Mr. Booth remains
largely unaffected. For them no drums have beat,
no trombones brayed ; no sanctified buffoonery, after
the model of the oration of the Friar in Wallen-
stein's camp dear to the readers of Schiller, has
tickled the ears of the groundlings on their behalf.
Sadly behind the great age of rowdy self-adver-
tisement in which their lot has fallen, they seem
not to have advanced one whit beyond John the

Baptist and the Apostles, 1800 years ago, in their
notions of the way in which the *metanoia*, the
change of mind of the ill-doer, is to be brought
about. Yet the new model was there, ready for
the imitation of those ancient savers of souls.
The ranting and roaring mystagogues of some of
the most venerable of Greek and Syrian cults also
had their processions and banners, their fifes and
cymbals and holy chants, their hierarchy of officers
to whom the art of making collections was not
wholly unknown; and who, as freely as their
modern imitators, promised an Elysian future to
contributory converts. The success of these
antique Salvation armies was enormous. Simon
Magus was quite as notorious a personage, and
probably had as strong a following as Mr. Booth.
Yet the Apostles, with their old-fashioned ways,
would not accept such a success as a satisfactory
sign of the Divine sanction, nor depart from their
own methods of leading the way to the higher life.

I deem it unessential to verify Mr. Booth's
statistics. The exact strength of the population
of the realm of misery, be it one, two, or three
millions, has nothing to do with the efficacy of
any means proposed for the highly desirable end
of reducing it to a *minimum*. The sole question
for consideration at present is whether the scheme,
keeping specially in view the spirit in which it
is to be worked, is likely to do more good than
harm.

Mr. Booth tells us, with commendable frankness, that "it is primarily and mainly for the sake of saving the soul that I seek the salvation of the body" (p. 45), which language, being interpreted, means that the propagation of the special Salvationist creed comes first, and the promotion of the physical, intellectual, and purely moral welfare of mankind second in his estimation. Men are to be made sober and industrious, mainly, that, as washed, shorn, and docile sheep, they may be driven into the narrow theological fold which Mr. Booth patronises. If they refuse to enter, for all their moral cleanliness, they will have to take their place among the goats as sinners, only less dirty than the rest.

I have been in the habit of thinking (and I believe the opinion is largely shared by reasonable men) that self-respect and thrift are the rungs of the ladder by which men may most surely climb out of the slough of despond of want; and I have regarded them as perhaps the most eminent of the practical virtues. That is not Mr. Booth's opinion. For him they are mere varnished sins —nothing better than "Pride re-baptised" (p. 46). Shutting his eyes to the necessary consequences of the struggle for life, the existence of which he accepts as fully as any Darwinian,[1] Mr. Booth tells men, whose evil case is one of those consequences, that envy is a corner-stone of our

[1] See p. 100.

competitive system. With thrift and self-respect
denounced as sin, with the suffering of starving
men referred to the sins of the capitalist, the
gospel according to Mr. Booth may save souls,
but it will hardly save society.

In estimating the social and political influence
which the Salvation Army is likely to exert, it
is important to reflect that the officers (pledged
to blind obedience to their " General ") are not
to confine themselves to the functions of mere
deacons and catechists (though, under a " General "
like Cyril, Alexandria knew to her cost what
even they could effect); they are to be " tribunes
of the people," who are to act as their gratuitous
legal advisers; and, when law is not sufficiently
effective, the whole force of the army is to obtain
what the said tribunes may conceive to be justice,
by the practice of ruthless intimidation. Society,
says Mr. Booth, needs " mothering "; and he sets
forth, with much complacency, a variety of
" cases," by which we may estimate the sort of
" mothering " to be expected at his parental
hands. Those who study the materials thus set
before them will, I think, be driven to the conclusion that the " mother " has already proved
herself a most unscrupulous meddler, even if
she has not fallen within reach of the arm of
the law.

Consider this " case." A, asserting herself to
have been seduced twice, " applied to our people.

We hunted up the man, followed him to the country, threatened him with public exposure, and forced from him the payment to his victim of £60 down, an allowance of £1 a week, and an insurance policy on his life for £450 in her favour" (p. 222).

Jedburgh justice this. "We" constitute ourselves prosecutor, judge, jury, sheriff's officer, all in one; "we" practise intimidation as deftly as if we were a branch of another League; and, under threat of exposure, "we" extort a tolerably heavy hush-money in payment of our silence.

Well, really, my poor moral sense is unable to distinguish these remarkable proceedings of the new popular tribunate from what, in French, is called *chantage* and, in plain English, blackmailing And when we consider that anybody, for any reason of jealousy, or personal spite, or party hatred, might be thus "hunted," "followed," "threatened," and financially squeezed or ruined, without a particle of legal investigation, at the will of a man whom the familiar charged with the inquisitorial business dare not hesitate to obey, surely it is not unreasonable to ask how far does the Salvation Army, in its "tribune of the people" aspect, differ from a Sicilian Mafia? I am no apologist of men guilty of the acts charged against the person who yet, I think, might be as fairly called a "victim," in this case, as his partner in wrong-doing. It is possible that, in so peculiar a case, Solomon himself might have been puzzled to apportion the relative moral delinquency of the parties. However that may be, the man was morally and legally bound to support his child, and any one would have been justified in helping the woman to her legal rights, and the man to the legal consequences (in which exposure is included) of his fault.

The action of the " General " of the Salvation Army in extorting the heavy fine he chose to impose as the price of his silence, however excellent his motives, appears to me to be as immoral as, I hope, it is illegal.

So much for the Salvation Army as a teacher of questionable ethics and of eccentric economics, as the legal adviser who recommends and practises the extraction of money by intimidation, as the fairy godmother who proposes to "mother" society, in a fashion which is not to my taste, however much it may commend itself to some of Mr. Booth's supporters.

I am, Sir, your obedient servant,

T. H. HUXLEY.

III

The " Times," December 11th, 1890

SIR,—When I first addressed you on the
subject of the projected operations of the
Salvation Army, all that I knew about that body
was derived from the study of Mr. Booth's book,
from common repute, and from occasional attention to the sayings and doings of his noisy
squadrons, with which my walks about London,
in past years, have made me familiar. I was
quite unaware of the existence of evidence respecting the present administration of the Salvation forces, which would have enabled me to act
upon the sagacious maxim of the American
humourist, " Don't prophesy unless you know."
The letter you were good enough to publish has
brought upon me a swarm of letters and pamphlets. Some favour me with abuse ; some
thoughtful correspondents warmly agree with me,
and then proceed to point out how much worthier
certain schemes of their own are of my friend's
support ; some send valuable encouragement, for
which I offer my hearty thanks, and ask them to
excuse any more special acknowledgment. But
that which I find most to the purpose, just now, is
the revelation made by some of the documents
which have reached me, of a fact of which I was

wholly ignorant—namely, that persons who have faithfully and zealously served in the Salvation Army, who express unchanged attachment to its original principles and practice, and who have been in close official relations with the "General," have publicly declared that the process of degradation of the organization into a mere engine of fanatical intolerance and personal ambition, which I declared was inevitable, has already set in and is making rapid progress.

It is out of the question, Sir, that I should occupy the columns of the "Times" with a detailed exposition and criticism of these *pièces justificatives* of my forecast. I say criticism, because the assertions of persons who have quitted any society must, in fairness, be taken with the caution that is required in the case of all *ex parte* statements of hostile witnesses. But it is, at any rate, a notable fact that there are parts of my first letter, indicating the inherent and necessary evil consequences of any such organization, which might serve for abstracts of portions of this evidence, long since printed and published under the public responsibility of the witnesses.

Let us ask the attention of your readers, in the first place, to "An ex-Captain's Experience of the Salvation Army," by J. J. R. Redstone, the genuineness of which is guaranteed by the preface (dated April 5th, 1888) which the Rev. Dr. Cunningham Geikie has supplied. Mr. Redstone's

story is well worth reading on its own account.
Told in simple, direct language such as John
Bunyan might have used, it permits no doubt
of the single-minded sincerity of the man, who
gave up everything to become an officer of the
Salvation Army, but, exhibiting a sad want of
that capacity for unhesitating and blind obedience
on which Mr. Booth lays so much stress, was
thrown aside, penniless—no, I am wrong, with
2s. 4d. for his last week's salary—to shift, with his
equally devoted wife, as he best might. I wish
I could induce intending contributors to Mr.
Booth's army chest to read Mr. Redstone's story.
I would particularly ask them to contrast the
pure simplicity of his plain tale with the artificial
pietism and slobbering unction of the letters
which Mr. Ballington Booth addresses to his
" dear boy " (a married man apparently older than
himself), so long as the said " dear boy " is facing
brickbats and starvation, as per order.

I confess that my opinion of the chiefs of the
Salvation Army has been so distinctly modified by
the perusal of this pamphlet that I am glad to be
relieved from the necessity of expressing it. It
will be much better that I should cite a few
sentences from the preface written by Dr.
Cunningham Geikie, who expresses warm admiration for the early and uncorrupted work of the
Salvation Army, and cannot possibly be accused
of prejudice against it on religious grounds :—

(1) "The Salvation Army 'is emphatically a family concern. Mr. Booth, senior, is General; one son is chief of the staff, and the remaining sons and daughters engross the other chief positions. It is Booth all over; indeed, like the sun in your eyes, you can see nothing else wherever you turn.' And, as Dr. Geikie shrewdly remarks, ' to be the head of a widely spread sect carries with it many advantages—not all exclusively spiritual.' "

(2) " Whoever becomes a Salvation officer is henceforth a slave, helplessly exposed to the caprice of his superiors."

" Mr. Redstone bore an excellent character both before he entered the army and when he left it. To join it, though a married man, he gave up a situation which he had held for five years, and he served Mr. Booth two years, working hard in most difficult posts. His one fault, Major Lawley tells us, was, that he was ' too straight '—that is, too honest, truthful, and manly—or, in other words, too real a Christian. Yet without trial, without formulated charges, on the strength of secret complaints which were never, apparently, tested, he was dismissed with less courtesy than most people would show a beggar—with 2s. 4d. for his last week's salary. If there be any mistake in this matter, I shall be glad to learn it."

(3) Dr. Geikie confirms, on the ground of information given confidentially by other officers,

Mr. Redstone's assertion that they are watched
and reported by spies from headquarters.

(4) Mr. Booth refuses to guarantee his officers
any fixed amount of salary. While he and his
family of high officials live in comfort, if not in
luxury, the pledged slaves whose devotion is the
foundation of any true success the Army has met
with often have "hardly food enough to sustain
life. One good fellow frankly told me that when
he had nothing he just went and begged."

At this point, it is proper that I should inter-
pose an apology for having hastily spoken of such
men as Francis of Assisi, even for purposes of
warning, in connection with Mr. Booth. Whatever may be thought of the wisdom of the plans
of the founders of the great monastic orders of
the middle ages, they took their full share of
suffering and privation, and never shirked in their
own persons the sacrifices they imposed on their
followers.

I have already expressed the opinion, that
whatever the ostensible purpose of the scheme
under discussion, one of its consequences will be
the setting up and endowment of a new Ranter-
Socialist sect. I may now add that another effect
will be—indeed, has been—to set up and endow
the Booth dynasty with unlimited control of the
physical, moral, and financial resources of the sect.
Mr. Booth is already a printer and publisher,
who, it is plainly declared, utilizes the officers of

the Army as agents for advertising and selling his publications; and some of them are so strongly impressed with the belief that active pushing of Mr. Booth's business is the best road to their master's favour, that when the public obstinately refuse to purchase his papers they buy them themselves and send the proceeds to headquarters. Mr. Booth is also a retail trader on a large scale, and the Dean of Wells has, most seasonably, drawn attention to the very notable banking project which he is trying to float. Any one who follows Dean Plumptre's clear exposition of the principles of this financial operation can have little doubt that, whether they are, or are not, adequate to the attainment of the first and second of Mr. Booth's ostensible objects, they may be trusted to effect a wide extension of any kingdom in which worldly possessions are of no value. We are, in fact, in sight of a financial catastrophe like that of Law a century ago. Only it is the poor who will suffer.

I have already occupied too much of your space, and yet I have drawn upon only one of the sources of information about the inner working of the Salvation Army at my disposition. Far graver charges than any here dealt with are publicly brought in the others.

I am, Sir, your obedient servant,

T. H. HUXLEY.

P.S.—I have just read Mr. Buchanan's letter in
the *Times* of to-day.   Mr. Buchanan is, I believe,
an imaginative writer.   I am not acquainted with
his works, but nothing in the way of fiction he
has yet achieved can well surpass his account of
my opinions and of the purport of my writings.

IV

The " Times," December 20th, 1890

SIR,—In discussing Mr. Booth's projects I have
hitherto left in the background a distinction
which must be kept well in sight by those who
wish to form a fair judgment of the influence, for
good or evil, of the Salvation Army.   Salvationism,
the work of " saving souls " by revivalist methods,
is one thing ; Boothism, the utilization of the
workers for the furtherance of Mr. Booth's
peculiar projects, is another.   Mr. Booth has
captured, and harnessed with sharp bits and
effectual blinkers, a multitude of ultra-Evangelical missionaries of the revivalist school who were
wandering at large.   It is this skilfully, if somewhat mercilessly, driven team which has dragged
the " General's " coach-load of projects into their
present position.

Looking, then, at the host of Salvationists proper, from the "captains" downwards (to whom, in my judgment, the family hierarchy stands in the relation of the Old Man of the Sea to Sinbad), as an independent entity, I desire to say that the evidence before me, whether hostile or friendly to the General and his schemes, is distinctly favourable to them. It exhibits them as, in the main, poor, uninstructed, not unfrequently fanatical, enthusiasts, the purity of whose lives, the sincerity of whose belief, and the cheerfulness of whose endurance of privation and rough usage, in what they consider a just cause, command sincere respect. For my part, though I conceive the corybantic method of soul-saving to be full of dangers, and though the theological speculations of these good people are to me wholly unacceptable, yet I believe that the evils which must follow in the track of such errors, as of all other errors, will be largely outweighed by the moral and social improvement of the people whom they convert. I would no more raise my voice against them (so long as they abstain from annoying their neighbours) than I would quarrel with a man, vigorously sweeping out a stye, on account of the shape of his broom, or because he made a great noise over his work. I have always had a strong faith in the principle of the injunction, "Thou shalt not muzzle the ox that treadeth out the corn." If a kingdom is worth a Mass, as

a great ruler said, surely the reign of clean living,
industry, and thrift is worth any quantity of
tambourines and eccentric doctrinal hypotheses.
All that I have hitherto said, and propose further
to say, is directed against Mr. Booth's extremely
clever, audacious, and hitherto successful attempt
to utilize the credit won by all this honest devotion and self-sacrifice for the purposes of his
socialistic autocracy.

I now propose to bring forward a little more
evidence as to how things really stand where Mr.
Booth's system has had a fair trial. I obtain it,
mainly, from a curious pamphlet, the title of
which runs: " The New Papacy. Behind the
Scenes in the Salvation Army," by an ex-Staff
Officer. " Make not my Father's house a house of
merchandise " (John ii. 16). 1889. Published
at Toronto, by A. Britnell. On the cover it is
stated that " This is the book which was burned
by the authorities of the Salvation Army." I
remind the reader, once more, that the statements
which I shall cite must be regarded as *ex parte ;*
all I can vouch for is that, on grounds of internal
evidence and from other concurrent testimony
respecting the ways of the Booth hierarchy, I
feel justified in using them.

This is the picture the writer draws of the army
in the early days of its invasion of the Dominion
of Canada :—

" Then, it will be remembered, it professed to be the humble handmaid of the existing churches; its professed object was the evangelization of the masses. It repudiated the idea of building up a separate religious body, and it denounced the practice of gathering together wealth and the accumulation of property. Men and women other than its own converts gathered around it and threw themselves heart and soul into the work, for the simple reason that it offered, as they supposed, a more extended and widely open field for evangelical effort. Ministers everywhere were invited and welcomed to its platforms, majors and colonels were few and far between, and the supremacy and power of the General were things unknown. . . . Care was taken to avoid anything like proselytism ; its converts were never coerced into joining its ranks. . . . In a word, the organization occupied the position of an auxiliary mission and recruiting agency for the various religious bodies. . . . The meetings were crowded, people professed conversion by the score, the public liberally supplied the means to carry on the work in their respective communities ; therefore every corps was wholly self-supporting, its officers were properly, if not luxuriously, cared for, the local expenditure was amply provided, and, under the supervision of the secretary, a local member, and the officer in charge, the funds were disbursed in the towns where they were collected, and the spirit of satisfaction and confidence was mutual all around " (pp. 4, 5).

Such was the army as the green tree. Now for the dry :—

" Those who have been daily conversant with the army's machinery are well aware how entirely and radically the whole system has changed, and how, from a band of devoted and disinterested workers, united in the bonds of zeal and charity for the good of their fellows, it has developed into a colossal and aggressive agency for the building up of a system and a sect, bound by rules and regulations altogether subversive of religious liberty and antagonistic to every (other ?) branch of Christian endeavour, and bound hand and foot to the will of one supreme head and ruler. . . . As the work has spread through the country, and as the area of its endeavours has enlarged, each leading position has been filled, one after the other, by individuals strangers to the country, totally ignorant of the sentiments and idiosyncrasies of the Canadian people, trained in one school under the teachings and dominance of a member of the Booth family, and out of whom every idea has been crushed, except that of unquestioning obedience to the General, and the absolute necessity of going forward to his bidding without hesitation or question " (p. 6).

"What is the result of all this? In the first place, whilst material prosperity has undoubtedly been attained, spirituality has been quenched, and, as an evangelical agency, the army has become almost a dead letter. . . . In seventy-five per cent. of its stations its officers suffer need and privation, chiefly on account of the heavy taxation that is placed upon them to maintain an imposing head-quarters and a large ornamental staff. The whole financial arrangements are carried on by a system of inflation and a hand-to-mouth extravagance and blindness as to future contingencies. Nearly all of its original workers and members have disappeared" (p. 7). "In reference to the religious bodies at large the army has become entirely antagonistic. Soldiers are forbidden by its rules to attend other places of worship without the permission of their officers. . . . Officers or soldiers who may conscientiously leave the service or the ranks are looked upon and often denounced publicly as backsliders. . . . Means of the most despicable description have been resorted to in order to starve them back to the service" (p. 8). "In its inner workings the army system is identical with Jesuitism. . . . That 'the end justifies the means,' if not openly taught, is as tacitly agreed as in that celebrated order" (p. 9).

Surely a bitter, overcharged, anonymous libel, is the reflection which will occur to many who read

these passages, especially the last. Well, I turn
to other evidence which, at any rate, is not anonymous. It is contained in a pamphlet entitled
" General Booth, the Family, and the Salvation
Army, showing its Rise, Progress, and Moral and
Spiritual Decline," by S. H. Hodges, LL.B., late
Major in the Army, and formerly private secretary
to General Booth (Manchester, 1890). I recommend potential contributors to Mr. Booth's wealth
to study this little work also. I have learned a
great deal from it. Among other interesting
novelties, it tells me that Mr. Booth has discovered " the necessity of a third step or blessing,
in the work of Salvation. He said to me one day,
' Hodges, you have only two barrels to your gun ;
I have three ' " (p. 31). And if Mr. Hodges's description of this third barrel is correct—" giving up
your conscience " and, " for God and the army,
stooping to do things which even honourable
worldly men would not consent to do " (p. 32)—it
is surely calculated to bring down a good many
things, the first principles of morality among
them.

Mr. Hodges gives some remarkable examples of
the army practice with the " General's " new rifle.
But I must refer the curious to his instructive
pamphlet. The position I am about to take up is
a serious one ; and I prefer to fortify it by the help
of evidence which, though some of it may be
anonymous, cannot be sneered away. And I shall

be believed, when I say that nothing but a sense of the great social danger of the spread of Boothism could induce me to revive a scandal, even though it is barely entitled to the benefit of the Statute of Limitations.

On the 7th of July, 1883, you, Sir, did the public a great service by writing a leading article on the notorious " Eagle " case, from which I take the following extract :—

" Mr. Justice Kay refused the application, but he was induced to refuse it by means which, as Mr. Justice Stephen justly remarked, were highly discreditable to Mr. Booth. Mr. Booth filed an affidavit which appears totally to have misled Mr. Justice Kay, as it would have misled any one who regarded it as a frank and honest statement by a professed teacher of religion."

When I addressed my first letter to you I had never so much as heard of the " Eagle " scandal. But I am thankful that my perception of the inevitable tendency of all religious autocracies towards evil was clear enough to bring about a provisional condemnation of Mr. Booth's schemes in my mind. Supposing that I had decided the other way, with what sort of feeling should I have faced my friend, when I had to confess that the money had passed into the absolute control of a person about the character of whose administra-

tion this concurrence of damnatory evidence was
already extant ?

I have nothing to say about Mr. Booth person-
ally, for I know nothing. On that subject, as on
several others, I profess myself an agnostic. But,
if he is, as he may be, a saint actuated by the
purest of motives, he is not the first saint who,
as you have said, has shown himself " in the
ardour of prosecuting a well-meant object " to
be capable of overlooking " the plain maxims of
every-day morality." If I were a Salvationist
soldier, I should cry with Othello, " Cassio, I love
thee ; but never more be officer of mine."

I am, Sir, your obedient servant,

T. H. HUXLEY.

V

The " Times," December 24*th,* 1890

SIR,—If I have any strong points, finance is
certainly not one of them. But the financial, or
rather fiscal, operations of the General of the
Salvation Army, as they are set forth and
exemplified in " The New Papacy," possess that
grand simplicity which is the mark of genius ;

and even I can comprehend them—or, to be more modest, I can portray them in such a manner that every lineament, however harsh, and every shade, however dark, can be verified by published evidence.

Suppose there is a thriving, expanding colonial town, and that, scattered among its artizans and labourers, there is a sprinkling of Methodists, or other such ultra-evangelical good people, doing their best, in a quiet way, to "save souls." Clearly, this is an outpost which it is desirable to capture. "We," therefore, take measures to get up a Salvation "boom" of the ordinary pattern. Enthusiasm is roused. A score or two of soldiers are enlisted into the ranks of the Salvation Army. "We" select the man who promises to serve our purposes best, make a "captain" of him, and put him in command of the "corps." He is very pleased and grateful; and indeed he ought to be. All he has done is that he has given up his trade; that he has promised to work at least nine hours a day in our service (none of your eight-hour nonsense for us) as collector, bookseller, general agent, and anything else we may order him to be. "We," on the other hand, guarantee him nothing whatever; to do so might weaken his faith and substitute worldly for spiritual ties between us. Knowing that, if he exerts himself in a right spirit, his labours will surely be blessed, we content ourselves with telling him that if, after all

expenses are paid and our demands are satisfied
each week, 25s. remains, he may take it. And, if
nothing remains, he may take that, and stay his
stomach with what the faithful may give him.
With a certain grim playfulness, we add that the
value of these contributions will be reckoned as
so much salary. So long as our "captain" is
successful, therefore, a beneficent spring of cash
trickles unseen into our treasury; when it begins
to dry up we say, "God bless you, dear boy," turn
him adrift (with or without 2s. 4d. in his pocket),
and put some other willing horse in the shafts.

The "General," I believe, proposes, among other
things, to do away with "sweating." May he not
as well set a good example by beginning at
home?

My little sketch, however, looks so like a
monstrous caricature that, after all, I must
produce the original from the pages of my
Canadian authority. He says that a "captain"
"has to pay 10 per cent. of all collections and
donations to the divisional fund for the support of
his divisional officer, who has also the privilege of
arranging for such special meetings as he shall
think fit, the proceeds of which he takes away for
the general needs of the division. Headquarters,
too, has the right to hold such special meetings
at the corps and send around such special at-
tractions as its wisdom sees fit, and to take away
the proceeds for the purposes it decides upon.

. . . He has to pay the rent of his building,
either to headquarters or a private individual ; he
has to send the whole collection of the afternoon
meeting of the first Sunday in the month to the
' Extension Fund ' at headquarters ; he has to pay
for the heating, lighting, and cleaning of his hall,
together with such necessary repairs as may be
needed ; he has to provide the food, lodging, and
clothing of his cadet, if he has one ; headquarters
taxes him with so many copies of the army papers
each week, for which he has to pay, sold or un-
sold ; and when he has done this, he may take
$6 (or $5, being a woman), or such proportion of
it as may be left, with which to clothe and feed
himself and to pay the rent and provide for the
heating and lighting of his quarters. If he has a
lieutenant he has to pay him $6 per week, or
such proportion of it as he himself gets, and share
the house expenses with him. Now, it will be
easily understood that at least 60 per cent. of the
stations in Canada the officer gets no money at
all, and he has to beg specially amongst his people
for his house-rent and food. There are few places
in the Dominion in which the soldiers do not find
their officers in all the food they need ; but it
must be remembered that the value of the food
so received has to be accounted for at headquarters
and entered upon the books of the corps as cash
received, the amount being deducted from any
moneys that the officer is able to take from the

week's collections. So that, no matter how much may be specially given, the officer cannot receive more than the value of $6 per week. The officer cannot collect any arrears of salary, as each week has to pay its own expenses; and if there is any surplus cash after all demands are met it must be sent to the 'war chest' at headquarters."
—" The New Papacy" (pp. 35, 36).

Evidently, Sir, "headquarters" has taken to heart the injunction about casting your bread upon the waters. It casts the crumb of a day or two's work of an emissary, and gets back any quantity of loaves of cash, so long as "captains" present themselves to be used up and replaced by new victims. What can be said of these devoted poor fellows except, *O sancta simplicitas !*

But it would be a great mistake to suppose that the money-gathering efficacy of Mr. Booth's fiscal agencies is exhausted by the foregoing enumeration of their regular operations. Consider the following edifying history of the "Rescue Home" in Toronto :—

" It is a fine building in the heart of the city; the lot cost $7,000, and a building was put up at a cost of $7,000 more, and there is a mortgage on it amounting to half the cost of the whole. The land to-day would probably fetch double its original price, and every year enhances its value . . . . . In the first five months of its

existence this institution received from the public
an income of $1,812 70c.; out of this $600 was
paid to headquarters for rent, $590 52c. was
spent upon the building in various ways, and the
balance of $622 18c. paid the salaries of the staff
and supported the inmates" (pp. 24, 25).

Said I not truly that Mr. Booth's fisc bears the
stamp of genius? Who else could have got the
public to buy him a "corner lot," put a building
upon it, pay all its working expenses: and then,
not content with paying him a heavy rent for the
use of the handsome present they had made him,
they say not a word against his mortgaging it to
half its value? And, so far as any one knows,
there is nothing to stop headquarters from selling
the whole estate to-morrow, and using the money
as the "General" may direct.

Once more listen to the author of "The New
Papacy," who affirms that "out of the funds
given by the Dominion for the evangelization of
the people by means of the Salvation Army, one
sixth had been spent in the extension of the
Kingdom of God, and the other five sixths had
been invested in valuable property, all handed
over to Mr. Booth and his heirs and assigns, as
we have already stated" (p. 26).

And this brings me to the last point upon
which I wish to touch. The answer to all
inquiries as to what has become of the enormous

personal and real estate which has been given over to Mr. Booth is that it is held "in trust." The supporters of Mr. Booth may feel justified in taking that statement "on trust." I do not. Anyhow, the more completely satisfactory this "trust" is, the less can any man who asks the public to put blind faith in his integrity and his wisdom object to acquaint them exactly with its provisions. Is the trust drawn up in favour of the Salvation Army? But what is the legal *status* of the Salvation Army? Have the soldiers any claim? Certainly not. Have the officers any legal interest in the "trust"? Surely not. The "General" has taken good care to insist on their renouncing all claims as a condition of their appointment. Thus, to all appearance, the army, as a legal person, is identical with Mr. Booth. And, in that case, any "trust" ostensibly for the benefit of the army is—what shall we say that is at once accurate and polite?

I conclude with these plain questions—Will Mr. Booth take counsel's opinion as to whether there is anything in such legal arrangements as he has at present made which prevents him from disposing of the wealth he has accumulated at his own will and pleasure? Will anybody be in a position to set either the civil or the criminal law in motion against him or his successors if he or they choose to spend every farthing in ways very different from those contemplated by the donors ?

I may add that a careful study of the terms of a "Declaration of Trust by William Booth in favour of the Christian Mission," made in 1878, has not enabled persons of much greater competence than myself to answer these questions satisfactorily.[1]

I am, Sir, your obedient servant,

T. H. Huxley.

On December 24th a letter appeared in the "Times" signed "J. S. Trotter," in which the following passages appear :—

"It seems a pity to put a damper on the spirits of those who agree with Professor Huxley in his denunciation of General Booth and all his works. May I give a few particulars as to the 'book' which was published in Canada? I had the pleasure of an interview with the author of a book written in Canada. The book was printed at Toronto, and two copies only struck off by the printers; one of these copies was stolen from the printer, and the quotation sent to you by Professor Huxley was inserted in the book, and is con-

---

[1] See *Preface* to this volume, pp. ix—xiii.

sequently a forgery. The book was published
without the consent and against the will of the
author.

" So the quotation is not only ' a bitter, overcharged anonymous libel,' as Professor Huxley
intimates, but a forgery as well. As to Mr.
Hodges, it seems to me to be simply trifling with
your readers to bring him in as an authority.
He was turned out of the army, out of kindness
taken on again, and again dismissed. If this had
happened to one of your staff, would his opinion
of the ' Times ' as a newspaper be taken for
gospel ? "

But in the " Times " of December 29th Mr. J.
S. Trotter writes :—

" I find I was mistaken in saying, in my letter
of Wednesday, to the ' Times ' that Mr. Hodges
was dismissed from the service of General Booth,
and regret any inconvenience the statement may
have caused to Mr. Hodges."

And on December 30th the " Times " published
a letter from Mr. Hodges in which he says that
Mr. Trotter's statements as they regard himself
" are the very reverse of truth. I was never
turned out of the Salvation Army. Nor, so far
as I was made acquainted with General Booth's
motives, was I taken on again out of kindness.

In order to rejoin the Salvation Army, I resigned the position of manager in a mill where I was in receipt of a salary of £250 per annum, with house-rent and one third of the profits. Instead of this Mr. Booth allowed me £2 per week and house-rent.

VI

The " Times," December 26*th,* 1890

SIR,—I am much obliged to Mr. J. S. Trotter for the letter which you published this morning. It furnishes evidence, which I much desired to possess on the following points :—

1. The author of " The New Papacy " is a responsible, trustworthy person ; otherwise Mr. Trotter would not speak of having had " the pleasure of an interview " with him.

2. After this responsible person had taken the trouble to write a pamphlet of sixty-four closely printed pages, some influence was brought to bear upon him, the effect of which was that he refused his consent to its publication. Mr. Trotter's excellent information will surely enable him to tell us what influence that was.

3. How does Mr. Trotter know that any passage

I have quoted is an interpolation? Does he possess that other copy of the "two" which alone, as he affirms, were printed?

4. If so, he will be able to say which of the passages I have cited is genuine and which is not; and whether the tenor of the whole uninterpolated copy differs in any important respect from that of the copy I have quoted.

It will be interesting to hear what Mr. J. S. Trotter has to say upon these points. But the really important thing which he has done is that he has testified, of his own knowledge, that the anonymous author of "The New Papacy" is no mere irresponsible libeller, but a person of whom even an ardent Salvationist has to speak with respect.

I am, Sir, your obedient servant,

T. H. HUXLEY.

[I may add that the unfortunate Mr. Trotter did me the further service of eliciting the letter from Mr. Hodges referred to on p. 277—which sufficiently establishes that gentleman's credit, and leads me to attach full weight to his evidence about the "third barrel."]

January 1891.

VII

The " Times," December 27*th,* 1890

SIR,—In making use of the only evidence of
the actual working of Mr. Booth's autocratic
government accessible to me, I was fully aware of
the slippery nature of the ground upon which I
was treading. For, as I pointed out in my first
letter, " no personal habit more surely degrades
the conscience and the intellect than blind and
unhesitating obedience to unlimited authority."
Now we have it, on Mr. Booth's own showing,
that every officer of his has undertaken to " obey
without questioning or gainsaying the orders from
headquarters."   And the possible relations of such
orders to honour and veracity are demonstrated
not only by the judicial deliverance on Mr. Booth's
affidavit in the " Eagle" case, which I have
already cited ; not only by Mr. Bramwell Booth's
admission before Mr. Justice Lopes that he had
stated what was " not quite correct " because he
had " promised Mr. Stead not to divulge " the facts
of the case (the " Times," November 4th, 1885);
but by the following passage in Mr. Hodges's
account of the reasons of his withdrawal from the
Salvation Army :—

" The General and Chief did not and could not

deny doing these things ; the only question was this, Was it right to practise this deception ? These points of difference were fully discussed between myself and the Chief of the Staff on my withdrawal, especially the Leamington incident, which was the one that finally drove me to decision. I had come to the conclusion, from the first, that they had acted as they supposed with a single eye to the good of God's cause, and had persuaded myself that the things were, as against the devil, right to be done, that as in battle one party captured and turned the enemy's own guns upon them, so, as they were fighting against the devil, it would be fair to use against him his weapons. And I wrote to this effect to the General" (p. 63).

Now, I do not wish to say anything needlessly harsh, but I ask any prudent man these questions. Could I, under these circumstances, trust any uncorroborated statement emanating from head-quarters, or made by the General's order ? Had I any reason to doubt the truth of Mr. Hodges's naïve confession of the corrupting influence of Mr. Booth's system ? And did it not behove me to pick my way carefully through the mass of statements before me, many of them due to people whose moral sense might, by possibility, have been as much blunted by the army discipline in the

use of the weapons of the devil as Mr. Hodges
affirms that his was ?

Therefore, in my third letter, I commenced my
illustrations of the practical working of Boothism
with the evidence of Mr. Redstone, fortified and
supplemented by that of a non-Salvationist, Dr.
Cunningham Geikie. That testimony has not
been challenged, and, until it is, I shall assume
that it cannot be. In my fourth letter, I cited a
definite statement by Mr. Hodges in evidence of
the Jesuitical principles of headquarters. What
sort of answer is it to tell us that Mr. Hodges
was dismissed the army ? A child might expect
that some such red herring would be drawn
across the trail ; and, in anticipation of the stale
trick, I added the strong *primâ facie* evidence of
the trustworthiness of my witness, in this particular, which is afforded by the "Eagle" case.
It was not until I wrote my fourth letter to you,
Sir—until the exploitation of the "captains" and
the Jesuitry of headquarters could be proved up
to the hilt—that I ventured to have recourse to
"The New Papacy." So far as the pamphlet
itself goes, this is an anonymous work ; and, for
sufficient reasons, I did not choose to go beyond
what was to be found between its covers. To
any one accustomed to deal with the facts of
evolution, the Boothism of "The New Papacy" was
merely the natural and necessary development
of the Boothism of Mr. Redstone's case and of the

"Eagle " case.   Therefore, I felt fully justified in
using it, at the same time carefully warning my
readers that it must be taken with due caution.

Mr. Trotter's useful letter admits that such a
book was written by a person with whom he had
the " pleasure of an interview," and that a version
of it (interpolated, according to his assertion) was
published against the will of the author.   Hence
I am justified in believing that there is a foundation of truth in certain statements, some of which
have long been in my possession, but which for
lack of Mr. Trotter's valuable corroboration I have
refrained from using.   The time is come when I
can set forth some of the heads of this information, with the request that Mr. Trotter, who
knows all about the business, will be so good as
to point out any error that there may be in them.
I am bound to suppose that his sole object, like
mine, is the elucidation of the truth, and to
assume his willingness to help me therein to the
best of his ability.

1. " The author of ' The New Papacy ' is a
Mr. Sumner, a person of perfect respectability, and
greatly esteemed in Toronto, who held a high
position in the Army.   When he left, a large
public meeting, presided over by a popular
Methodist minister, passed a vote of sympathy
with him."

Is this true or false ?

2. " On Saturday last, about noon, Mr. Sumner, the author of the book, and Mr. Fred Perry, the Salvation Army printer, accompanied by a lawyer, went down to Messrs. Imrie and Graham's establishment, and asked for all the manuscript, stereotype plates, &c., of the book.   Mr. Sumner explained that the book had been sold to the Army, and, on a cheque for the amount due being given, the printing material was delivered up."

Did these paragraphs appear in the "Toronto Telegram" of April 24th, 1889, or did they not ?   Are the statements they contain true or false ?

3. " Public interest in the fate or probable outcome of that mysterious book called ' The New Papacy ; or, Behind the Scenes in the Salvation Army,' continues unabated, though the line of proceedings by the publisher and his solicitor, Mr. Smoke, of Watson, Thorne, Smoke, and Masten, has not been altered since yesterday.   The book, no doubt, will be issued in some form.   So far as known, only one complete copy remains, and the whereabouts of this is a secret which will be profoundly kept.   It is safe to say that if the Commissioner kept on guessing until the next anniversary, he would not strike the secluded

location of the one volume among five thousand which escaped, when he and his assistant, Mr. Fred Perry, believed they had cast every vestige of the forbidden work into the fiery furnace. On Tuesday last, when the discovery was made that a copy of 'The New Papacy' was in existence, Publisher Britnell, of Yonge Street, was at once the suspected holder, and in a short time his book-store was the resort of army agents sent to reconnoitre" ("Toronto News," April 28th, 1889).

Is this a forgery, or is it not? Is it in substance true or false?

When Mr. Trotter has answered these inquiries categorically, we may proceed to discuss the question of interpolations in Mr. Sumner's book.

I am, Sir, your obedient servant,

T. H. HUXLEY.

[On the 26th of December a letter, signed J. T. Cunningham, late Fellow of University College, Oxford, called forth the following commentary.]

VIII

The " Times," December 29*th,* 1890

SIR,—If Mr. Cunningham doubts the efficacy of the struggle for existence, as a factor in social conditions, he should find fault with Mr. Booth and not with me.

" I am labouring under no delusion as to the possibility of inaugurating the millennium by my social specific. In the struggle of life the weakest will go to the wall, and there are so many weak. The fittest in tooth and claw will survive. All that we can do is to soften the lot of the unfit, and make their suffering less horrible than it is at present " ("In Darkest England," p. 44).

That is what Mr. Cunningham would have found if he had read Mr. Booth's book with attention. And, if he will bestow equal pains on my second letter, he will discover that he has interpolated the word " wilfully " in his statement of my " argument," which runs thus : " Shutting his eyes to the necessary consequences of the struggle for life, the existence of which he admits as fully as any Darwinian, Mr. Booth tells men whose evil case is one of those consequences that envy is a corner-stone of our competitive system."

Mr. Cunningham's physiological studies will have informed him that the process of "shutting the eyes," in the literal sense of the words, is not always wilful; and I propose to illustrate, by the crucial instance his own letter furnishes, that the "shutting of the eyes" of the mind to the obvious consequences of accepted propositions may also be involuntary. At least, I hope so.

1. "Sooner or later," says Mr. Cunningham, "the population problem will block the way once more." What does this mean, except that multiplication, excessive in relation to the contemporaneous means of support, will create a severe competition for those means? And this seems to me to be a pretty accurate "reflection of the conceptions of Malthus" and the other poor benighted folks of a past generation at whom Mr. Cunningham sneers.

2. By way of leaving no doubt upon this subject, Mr. Cunningham further tells us, "The struggle for existence is always going on, of course; let us thank Darwin for making us realize it." It is pleasant to meet with a little gratitude to Darwin among the *epigoni* who are squabbling over the heritage he conquered for them, but Mr. Cunningham's personal expression of that feeling is hasty. For it is obvious that he has not "realized" the significance of Darwin's teaching —indeed, I fail to discover in Mr. Cunningham's letter any sign that he has even "realized" what

he would be at.   If the " struggle for existence is
always going on"; and if, as I suppose will be
granted, industrial competition is one phase of
that struggle, I fail to see how my conclusion that
it is sheer wickedness to tell ignorant men that
" envy " is a corner-stone of competition can be
disputed.

Mr. Cunningham has followed the lead of that
polished and instructed person, Mr. Ben Tillett,
in rebuking me for (as the associates say) attack-
ing Mr. Booth's personal character.   Of course,
when I was writing, I did not doubt that this
very handy, though not too clean, weapon would
be used by one or other of Mr. Booth's supporters.
And my action was finally decided by the following considerations: I happen to be a member of
one of the largest life insurance societies.   There
is a vacancy in the directory at present, for which
half a dozen gentlemen are candidates.   Now, I
said to myself, supposing that one of these gentlemen (whose pardon I humbly beg for starting the
hypothesis), say Mr. A., in his administrative
capacity and as a man of business, has been the
subject of such observations as a Judge on the
Bench bestowed upon Mr. Booth, is he a person for
whom I can properly vote ?   And, if I find, when
I go to the meeting of the policy-holders, that
most of them know nothing of this and other
evidences of what, by the mildest judgment, must
be termed Mr. A.'s unfitness for administrative

responsibilities, am I to let them remain in their
ignorance ? I leave the answer and its application to men of sense and integrity.

The mention of Mr. Cunningham's ally reminds
me that I have omitted to thank Mr. Tillett for
his very useful and instructive letter; and I
hasten to repair a neglect which I assure Mr.
Tillett was more apparent than real. Mr. Tillett's
letter is dated December 20th. On the 21st
the following pregnant (however unconscious)
commentary upon it appeared in " Reynolds's
Newspaper " :—

" I have always maintained that the Salvation
Army is one of the mightiest Socialistic agencies
in the country; and now Professor Huxley comes
in to confirm that view. How could it be otherwise ? The fantastic religious side of Salvationism
will disappear in the course of time, and what
will be left ? A large number of men and women
who have been organized, disciplined, and taught
to look for something better than their present
condition, and who have become public speakers
and not afraid of ridicule. There you have the
raw materials for a Socialist army."

Mr. Ben Tillett evidently knows Latin enough
to construe *proximus ardet*.

I trust that the public will not allow themselves
to be led away by the false issues which are

dangled before them.   A man really may love
his fellow-men; cherish any form of Christianity
he pleases; and hold not only that Darwinism
is "tottering to its fall," but, if he pleases, the
equally sane belief that it never existed; and yet
may feel it his duty to oppose, to the best of his
capacity, despotic Socialism in all its forms, and,
more particularly, in its Boothian disguise.

I am, Sir, your obedient servant,

T. H. HUXLEY.

[Persons who have not had the advantage of a
classical education might fairly complain of my
use of the word *epigoni*.   To say truth, I had
been reading Droysen's "Geschichte des Hellen-
ismus," and the familiar historical title slipped
out unawares.   In replying to me, however, the
late "Fellow of University College," Oxford,
declares he had to look the word out in a
Lexicon.   I commend the fact to the notice of
the combatants over the desirability of retaining
the present compulsory modicum of Greek in our
Universities.]

IX

The " Times," December 30*th*, 1890

SIR,—I am much obliged to Messrs. Ranger,
Burton, and Matthews for their prompt answer to
my questions. I presume it applies to all money
collected by the agency of the Salvation Army,
though not specifically given for the purposes
of the "Christian Mission" named in the deed
of 1878; to all sums raised by mortgage upon
houses and land so given; and, further, to funds
subscribed for Mr. Booth's various projects, which
have no apparent reference to the objects of
the "Christian Mission," as defined in the deed.
Otherwise, to use a phrase which has become
classical, "it does not assist us much." But I
must leave these points to persons learned in the
law.

And, indeed, with many thanks to you, Sir, for
the amount of valuable space which you have
allowed me to occupy, I now propose to leave the
whole subject. My sole purpose in embarking
upon an enterprise which was extremely distasteful to me was to prevent the skilful
"General," or rather "Generals," who devised
the plan of campaign from sweeping all before
them with a rush. I found the pass already held
by such stout defenders as Mr. Loch and the

Dean of Wells, and, with your powerful help, we have given time for the reinforcements, sure to be sent by the abundant, though somewhat slowly acting, common sense of our countrymen, to come up.

I can no longer be useful, and I return to more congenial occupations.

I am, Sir, your obedient servant,

T. H. HUXLEY.

The following letter appeared in the "Times" of January 2nd, 1891 :—

"DEAR MR. TILLETT,—I have not had patience to read Professor Huxley's letters. The existence of hunger, nakedness, misery, 'death from insufficient food,' even of starvation, is certain, and no agency as yet reaches it. How can any man hinder or discourage the giving of food or help? Why is the house called a workhouse? Because it is for those who cannot work? No, because it was the house to give work or bread. The very name is an argument. I am very sure what Our Lord and His Apostles would do if they were in London. Let us be thankful even to have a will to do the same.

"Yours faithfully,

"HENRY E. CARD. MANNING."

X

The "Times," January 3rd, 1891

SIR,—In my old favourite, " The Arabian
Nights," the motive of the whole series of delightful narratives is that the sultan, who refuses to
attend to reason, can be got to listen to a story.
May I try whether Cardinal Manning is to be
reached in the same way ? When I was attending the meeting of the British Association in
Belfast nearly forty years ago, I had promised to
breakfast with the eminent scholar Dr. Hincks.
Having been up very late the previous night, I
was behind time ; so, hailing an outside car, I said
to the driver as I jumped on, " Now drive fast, I
am in a hurry." Whereupon he whipped up his
horse and set off at a hand-gallop. Nearly jerked
off my seat, I shouted, " My good friend, do you
know where I want to go ? " " No, yer honner,"
said the driver, " but, any way, I am driving fast."
I have never forgotten this object-lesson in the
dangers of ill-regulated enthusiasm. We are all invited to jump on to the Salvation Army car, which
Mr. Booth is undoubtedly driving very fast. Some
of us have a firm conviction, not only that he is
taking a very different direction from that in
which we wish to go, but that, before long, car
and driver will come to grief. Are we to accept

the invitation, even at the bidding of the eminent person who appears to think himself entitled to pledge the credit of " Our Lord and His Apostles " in favour of Boothism ?

I am, Sir, your obedient servant,

T. H. HUXLEY.

XI

The " Times," January 13*th,* 1891

SIR,—A letter from Mr. Booth-Clibborn, dated January 3rd, appeared in the " Times " of yester-day. This elaborate document occupies three columns of small print—space enough, assuredly, for an effectual reply to the seven letters of mine to which the writer refers, if any such were forthcoming. Mr. Booth-Clibborn signs himself " Commissioner of the Salvation Army for France and Switzerland," but he says that he accepts my " challenge " without the knowledge of his chiefs. Considering the self-damaging character of his letter, it was, perhaps, hardly necessary to make that statement.

Mr. " Commissioner " Booth-Clibborn speaks of my " challenge." I presume that he refers to my

request for information about the authorship and
fate of " The New Papacy," in the letter published
in the " Times " on December 27th, 1890. The
" Commissioner " deals with this matter in paragraph No. 4 of his letter; and I observe, with no
little satisfaction, that he does not venture to con-
trovert any one of the statements of my witnesses.
He tacitly admits that the author of " The New
Papacy " was a person " greatly esteemed in
Toronto," and that he held " a high position in the
army "; further, that the Canadian " Commissioner " thought it worth while to pay the printer's
bill, in order that the copies already printed off
might be destroyed and the pamphlet effectually
suppressed. Thus the essential facts of the case
are admitted and established beyond question.

How does Mr. Booth-Clibborn try to explain
them away ?

" Mr. Sumner, who wrote the little book in a
hot fit, soon regretted it (as any man would do
whose conscience showed him in a calmer moment,
when his ' respectability ' returned with his repentance, that he had grossly misrepresented), and just
before it appeared offered to order its suppression
if the army would pay the costs already incurred,
and which he was unable to bear."

" The New Papacy " fills sixty closely printed
duodecimo pages. It is carefully written, and for
the most part in studiously moderate language;

moreover, it contains many precise details and figures, the ascertainment of which must have taken much time and trouble. Yet, forsooth, it was written in " a hot fit."

I sincerely hope, for the sake of his own credit, that Mr. " Commissioner " Booth-Clibborn does not know as much about this melancholy business as I do. My hands are unfortunately tied, and I am not at liberty to use all the information in my possession. I must content myself with quoting the following passage from the preface to " The New Papacy " :—

" It has not been without considerable thought and a good deal of urging that the following pages have been given to the public. But though we would have shrunk from a labour so distasteful, and have gladly avoided a notoriety anything but pleasant to the feelings, or conducive to our material welfare, we have felt that in the interests of the benevolent public, in the interests of religion, in the interests of a band of devoted men and women whose personal ends are being defeated, and the fruit of whose labour is being destroyed, and, above all, in the interests of that future which lies before the Salvation Army itself, if purged and purified in its executive and returned to its original position in the ranks of Canadian Christian effort, it is no more than our duty to throw such light as we are able upon its true inwardness, and with that object and for the furtherance of those ends we offer our pages to the public view."

The preface is dated April 1889. According to the statement in the "Toronto Telegram," which Mr. " Commissioner " Booth-Clibborn does not dare to dispute, his Canadian fellow-" Commissioner " bought and destroyed the whole edition of " The New Papacy " about the end of the third week in April. It is clear that the writer of the paragraph quoted from the preface was well out of a " hot fit," if he had ever been in one, while he had not entered on the stage of repentance within three weeks of that time. Mr. " Commissioner " Booth-Clibborn's scandalous insinuations that Mr. Sumner was bribed by " a few sovereigns," and that he was " bought off," in the face of his own admission that Mr. Sumner " offered to order its suppression if the army would pay the costs already incurred, and which he was unable to bear," is a crucial example of that Jesuitry with which the officials of the army have been so frequently charged.

Mr. " Commissioner " Booth-Clibborn says that when " London headquarters heard of the affair, it disapproved of the action of the Commissioner." That circumstance indicates that headquarters is not wholly devoid of intelligence; but it has nothing to do with the value of Mr. Sumner's evidence, which is all I am concerned about. Very likely London headquarters will disapprove

of its French " Commissioner's " present action.
But what then ?   The upshot of all this is that
Mr. Booth-Clibborn has made as great a blunder
as simple Mr. Trotter did.   The pair of Balaams
greatly desired to curse, but have been compelled
to bless.   They have, between them, completely
justified my reliance on Mr. Sumner as a perfectly
trustworthy witness ; and neither of them has
dared to challenge the accuracy of one solitary
statement made by that worthy gentleman, whose
full story I hope some day or other to see set
before the public.   Then the true causes of his
action will be made known.

Paragraph 2 of the " Commissioner's " letter
says many things, but not much about Mr. Hodges.
The columns of the " Times " recently showed
that Mr. Hodges was able to compel an apology
from Mr. Trotter.   I leave it to him to deal with
the " Commissioner."

As to the " Eagle " case, treated of in paragraph
No. 3, a gentleman well versed in the law, who
was in court during the hearing of the appeal,
has assured me that the argument was purely
technical ; that the facts were very slightly gone
into ; and that, so far as he knows, no dissenting
comment was made on the strictures of the Judge
before whom the case first came.   Moreover, in
the judgment of the Master of the Rolls, fully recorded in the " Times " of February 14th, 1884,
the following passages occur :—

"The case had been heard by a learned Judge, who had exercised his discretion upon it, and the Court would not interfere with his discretion unless they could see that he was wrong. The learned Judge had taken a strong view of the conduct of the defendant, but nevertheless had said that he would have given relief if he could have seen how far protection and compensation could be given. And if this Court differed from him in that view, and could give relief without forfeiture, they would be acting on his own principle in doing so. Certain suggestions had been made with that view, and the Court had to consider the case under all the circumstances. . . . He himself (the Master of the Rolls) considered that it was probable the defendant, with his principles, had intended to destroy the property as a public-house, and that it was not right thus to take property under a covenant to keep it up as a public-house, intending to destroy it as such. He did not, however, think this was enough to deprive him of all relief. . . . The defendant could only expect severe terms."

Yet, Sir, Mr. "Commissioner" Booth-Clibborn, this high official of the Salvation Army, has the audacity to tell the public that if I had made inquiries I should have found that "in the Court of Appeal the Judge reversed the decision of his predecessor as regards seven eighths of the property, and the General was declared to have acted

all along with straightforwardness and good
faith."

But the nature of Mr. "Commissioner" Booth-
Clibborn's conceptions of straightforwardness and
good faith is so marvellously illustrated by the
portions of his letter with which I have dealt
that I doubt not his statements are quite up to
the level of the "Army" Regulations and Instruc-
tions in regard to those cardinal virtues. As I
pointed out must be the case, the slave is subdued
to that he works in.

For myself, I must confess that the process of
wading through Mr. "Commissioner's" verbose
and clumsy pleadings has given me a "hot fit,"
which, I undertake to say, will be followed by not
so much as a passing shiver of repentance. And
it is under the influence of the genial warmth
diffused through the frame, on one of those rare
occasions when one may be "angry and sin not,"
that I infringe my resolution to trouble you with
no more letters. On reflection, I am convinced
that it is undesirable that the public should be
misled, for even a few days, by misrepresentations
so serious.

I am copiously abused for speaking of the
Jesuitical methods of the superior officials of the
Salvation Army. But the following facts have
not been, and, I believe, cannot be, denied :—

1. Mr. Booth's conduct in the "Eagle" case
has been censured by two of the Judges.

2. Mr. Bramwell Booth admitted before Mr. Justice Lopes that he had made an untrue statement because of a promise he had made to Mr. Stead.[1]

And I have just proved that Mr. "Commissioner" Booth-Clibborn asserts the exact contrary of that which your report of the judgment of the Master of the Rolls tells us that distinguished judge said.

Under these circumstances, I think that my politeness in applying no harder adjective than "Jesuitical" to these proceedings is not properly appreciated.

I am, Sir, your obedient servant,

T. H. HUXLEY.

XII

The "Times," January 22*nd,* 1891

SIR,—I think that your readers will be interested in the accompanying opinion, written in consultation with an eminent Chancery Queen's Counsel, with which I have been favoured. It will be observed that this important legal de-

---

[1] This statement has been disputed, but not yet publicly. (See p. 305.)

liverance justifies much stronger language than
any which I have applied to the only security (?)
for the proper administration of the funds in Mr.
Booth's hands which appears to be in existence.

I am, Sir, your obedient servant,

T. H. HUXLEY.

1, DR. JOHNSON'S BUILDINGS, TEMPLE, E.C.,
January 14, 1891.

MR. BOOTH'S DECLARATION OF TRUST DEED, 1878.

"I am of opinion, subject to the question
whether there may be any provision in the
Charitable Trusts Acts which can be made available for enforcing some scheme for the appropriation of the property, and with regard to the real
and leasehold properties whether the conveyances
and leases are not altogether void, as frauds on
the Mortmain Acts, that nothing can be done to
control or to interfere with Booth in the disposition or application of the properties or moneys
purported to be affected by the deed.

"As to the properties vested in Booth himself,
it appears to me that such are placed absolutely
under his power and control both as to the disposal and application thereof, and that there are no
trusts for any specific purposes declared which

could be enforced, and that there are no defined persons nor classes of persons who can claim to be entitled to the benefit of them, or at whose instance they could be enforced by any legal process.

" As to the properties (if any) vested in trustees appointed by Booth, it appears to me that the only person who has a *locus standi* to enforce these trusts is Booth himself, and that he would have absolute power over the trusts and the property, and might deal with the property as he pleased, and that, as in the former case, nothing could be done in the way of enforcing any trusts against him.

" As to the moneys contributed or raised by mortgage for the general purposes of the mission, it appears to me that Booth may expend them as he pleases, without being subject to any legal control, and that he cannot even be compelled to publish any balance-sheets.

" Whether there are any provisions in the Charitable Trusts Acts which could be made available for enforcing some scheme for the application of the property or funds is a question to which I should require to give a closer consideration should it become necessary to go into it ; but at present, after perusing these Acts, and especially 16 and 17 Vict. c. 137 and 18 and 19 Vict. c. 124, I cannot see how they could be made applicable to the trusts as declared in this deed.

" As to the Mortmain Acts, the matter is clearly
charitable, and unless in the conveyances and
leases to Booth, or to the trustees (if any) named
by him, all the provisions of the Acts have been
complied with, and the deeds have been enrolled
under the Acts, they would be void. It is probable, however, that every conveyance and lease has
been taken without disclosing any charitable trust,
for the purpose of preventing it from being void
on the face of it. It is to be noted that the deed
is a mere deed poll by Booth himself, without any
other party to it, who, as a contracting party,
would have a right to enforce it.

" Whether there are any objects of the trust I
cannot say. If there is, as the recital indicates, a
society of enrolled members called ' The Christian
Mission,' those members would be objects of the
trust, but then, it appears to me, Booth has entire
control and determination of the application.
And, as to the trusts enuring for the benefit of
the ' Salvation Army,' I am not aware what is the
constitution of the ' Salvation Army,' but there is
no reference whatever to any such body in the
deed. I have understood the army as being
merely the missionaries, and not the society of
worshippers.

" If there is no Christian Mission Society of
enrolled members, then there are no objects of the
trust. The trusts are purely religious, and trading
is entirely beyond its purposes. Booth can ' give

away ' the property, simply because there is no one
who has any right to prevent his doing so.

" ERNEST HATTON."

It is probably my want of legal knowledge
which prevents me from appreciating the value of
the professed corrections of Mr. Hatton's opinion
contained in the letters of Messrs. Ranger, Burton,
and Matthews, " Times," January 28th and 29th,
1891.

The note on page 301 refers to a correspondence,
incomplete at the time fixed for the publication of
my pamphlet, the nature of which is sufficiently
indicated by the subjoined extracts from Mr.
Stead's letter in the " Times " of January 20th,
and from my reply in the " Times " of January
24th. Referring to the paragraphs numbered 1,
2, at the end of my letter XI., Mr. Stead says :—

" On reading this, I at once wrote to Professor
Huxley, stating that, as he had mentioned my
name, I was justified in intervening to explain
that, so far as the second count in his indictment
went—for the Eagle dispute is no concern of
mine—he had been misled by an error in the
reports of the case which appeared in the daily

papers of November 4, 1885. I have his reply
to-day, saying that I had better write to you
direct. May I ask you, then, seeing that my
name has been brought into the affair, to state
that, as I was in the dock when Mr. Bramwell
Booth was in the witness-box, I am in a position
to give the most unqualified denial to the statement as to the alleged admission on his part of
falsehood ? Nothing was heard in Court of any
such admission. Neither the prosecuting counsel
nor the Judge who tried the case ever referred to
it, although it would obviously have had a direct
bearing on the credit of the witness; and the
jury, by acquitting Mr. Bramwell Booth, showed
that they believed him to be a witness of truth.
But fortunately the facts can be verified beyond
all gainsaying by a reference to the official short-
hand-writer's report of the evidence. During the
hearing of the case for the prosecution, Inspector
Borner was interrupted by the Judge, who
said :—

" ' I want to ask you a question. During the
whole of that conversation, did Booth in any way
suggest that that child had been sold ? ' Borner
replied :—
" ' Not at that interview, my Lord.'

" It was to this that Mr. Bramwell Booth
referred when, after examination, cross-examina-

tion, and re-examination, during which no suggestion had been made that he had ever made the untrue statement now alleged against him, he asked and received leave from the Judge to make the following explanation, which I quote from the official report :—

" ' Will you allow me to explain a matter mentioned yesterday in reference to a question asked by your Lordship some days ago with respect to one matter connected with my conduct ? Your Lordship asked, I think it was Inspector Borner, whether I had said to him at either of our interviews that the child was sold by her parents, and he replied " No." That is quite correct; I did not say so to him, and what I wish to say now is that I had been specially requested by Mr. Stead, and had given him a promise, that I would not under any circumstances divulge the fact of that sale to any person which would make it at all probable that any trouble would be brought upon the persons who had taken part in this investigation.' (Central Criminal Court Reports, Vol. CII., part 612, pp. 1,035-6.)

" In the daily papers of the following day this statement was misreported as follows :—

" ' I wish to explain, in regard to your Lordship's condemnation of my having said " No " to

Inspector Borner when he asked me whether the
child had been sold by her parents—the reason
why I stated what was not correct was that I had
promised Mr. Stead not to divulge the fact of the
sale to any person which would make it probable
that any trouble should be brought on persons
taking part in this proceeding.'

"Hence the mistake into which Professor
Huxley has unwittingly fallen.

"I may add that, so far from the statement
never having been challenged for five years, it
was denounced as 'a remarkably striking lie' in
the 'War Cry' of November 14th, and again the
same official organ of the Salvation Army of
November 18th specifically adduced this mis-
report as an instance of 'the most disgraceful
way' in which the reports of the trial were garbled
by some of the papers. What, then, becomes of
one of the two main pillars of Professor Huxley's
argument?"

In my reply, I point out that, on the 10th of
January, Mr. Stead addressed to me a letter,
which commences thus : "I see in the 'Times'
of this morning that you are about to republish
your letters on Booth's book."
I replied to this letter on the 12th of
January :—

"DEAR MR. STEAD,—I charge Mr. Bramwell
Booth with nothing. I simply quote the 'Times'
report, the accuracy of which, so far as I know,
has never been challenged by Mr. Booth. I say
I quote the 'Times' and not Mr. Hodges,[1]
because I took some pains about the verification
of Mr. Hodges's citation.

"I should have thought it rather appertained
to Mr. Bramwell Booth to contradict a statement
which refers, not to what you heard, but to what
he said. However, I am the last person to wish
to give circulation to a story which may not be
quite correct; and I will take care, if you have no
objection (your letter is marked 'private'), to
make public as much of your letter as relates to
the point to which you have called my attention.

"I am, yours very faithfully,

"T. H. HUXLEY."

To this Mr. Stead answered, under date of
January 13th, 1891 :—

"DEAR PROFESSOR HUXLEY,—I thank you for
your letter of the 12th inst. I am quite sure you
would not wish to do any injustice in this matter.
But, instead of publishing any extract from my
letter, might I ask you to read the passage as it

[1] This is a slip of the pen. Mr. Hodges had nothing to do
with the citation of which I made use.

appears in the verbatim report of the trial which
was printed day by day, and used by counsel on
both sides, and by the Judge during the case ?
I had hoped to have got you a copy to-day, but
find that I was too late. I shall have it first
thing to-morrow morning. You will find that it
is quite clear, and conclusively disposes of the
alleged admission of untruthfulness. Again
thanking you for your courtesy,

" I am, yours faithfully,

" W. T. STEAD."

Thus it appears that the letter which Mr. Stead
wrote to me on the 13th of January does not
contain one word of that which he says it contains, in the statement which appears in the
" Times " to-day. Moreover, the letter of mine to
which Mr. Stead refers in his first communication
to me is not the letter which appeared on the
13th, as he states, but that which you published
on December 27th, 1890. Therefore, it is not
true that Mr. Stead wrote " at once." On the
contrary, he allowed nearly a fortnight to elapse
before he addressed me on the 10th of January
1891. Furthermore, Mr. Stead suppresses the
fact that, since the 13th of January, he has had
in his possession my offer to publish his version of
the story ; and he leads the reader to suppose that
my only answer was that he " had better write to

you direct." All the while, Mr. Stead knows
perfectly well that I was withheld from making
public use of his letter of the 10th by nothing but
my scruples about using a document which was
marked " private " ; and that he did not give me
leave to quote his letter of the 10th of January
until after he had written that which appeared
yesterday.

And I add :—

As to the subject-matter of Mr. Stead's letter,
the point which he wishes to prove appears to be
this—that Mr. Bramwell Booth did not make a
false statement, but that he withheld from the
officers of justice, pursuing a most serious criminal
inquiry, a fact of grave importance, which lay
within his own knowledge. And this because he
had promised Mr. Stead to keep the fact secret.
In short, Mr. Bramwell Booth did not say what
was wrong ; but he did what was wrong.

I will take care to give every weight to the
correction. Most people, I think, will consider
that one of the " main pillars of my argument,"
as Mr. Stead is pleased to call them, has become
very much strengthened.

LEGAL OPINIONS RESPECTING
"GENERAL" BOOTH'S ACTS.

In referring to the course of action adopted by
"General" Booth and Mr. Bramwell Booth in
respect of their legal obligations to other persons,
or to the criminal and civil law, I have been as
careful as I was bound to be, to put any difficulties suggested by mere lay common-sense in
an interrogative or merely doubtful form; and to
confine myself, for any positive expressions, to
citations from published declarations of the
judges before whom the acts of "General" Booth
came; from reports of the Law Courts; and from
the deliberate opinions of legal experts. I have
now some further remarks to make on these
topics.

I. The observations at p. 305 express, with
due reserve, the impression which the counsel's
opinions, quoted by "General" Booth's solicitors,
made on my mind. They were written and sent
to the printer before I saw the letter from a
"Barrister *not* Practising on the Common Law
Side," and those from Messrs. Clarke and Calkin
and Mr. George Kebbell, which appeared in the
"Times" of February 3rd and 4th.

These letters fully bear out the conclusion
which I had formed, but which it would have

been presumptuous on my part to express, that
the opinions cited by "General" Booth's solicitors
were like the famous broken tea-cups "wisely
ranged for show"; and that, as Messrs. Clarke
and Calkin say, they "do not at all meet the
main points on which Mr. Hatton advised." I
do not think that any one who reads attentively
the able letter of "A Barrister *not* Practising on
the Common Law Side" will arrive at any other
conclusion; or who will not share the very natural
desire of Mr. Kebbell to be provided with clear
and intelligible answers to the following inquiries :—

(1) Does the trust deed by its operation
empower any one legally to call upon Mr. Booth
to account for the application of the funds?

(2) In the event of the funds not being properly
accounted for, is any one, and, if so, who, in a
position to institute civil or criminal proceedings
against any one, and whom, in respect of such
refusal or neglect to account?

(3) In the event of the proceedings, civil or
criminal, failing to obtain restitution of misapplied
funds, is or are any other person or persons liable
to make good the loss?

On December 24th, 1890, a letter of mine
appeared in the "Times" (No. V. above) in which
I put questions of the same import, and asked
Mr. Booth if he would not be so good as to take
counsel's opinion on the "trusts" of which so

much has been heard and so little seen, not as
they stood in 1878, or in 1888, but as they stand
now? Six weeks have elapsed, and I wait for a
reply.

It is true that Dr. Greenwood has been author-
ized by Mr. Booth to publish what he calls a
"Rough outline of the intended Trust Deed"
("General Booth and His Critics," p. 120), but
unfortunately we are especially told that it "*does
not profess to be an absolutely accurate analysis.*"
Under these circumstances I am afraid that
neither lawyers nor laymen of moderate intelligence will pay much attention to the assertion,
that "*it gives a fair idea of the general effect of
the draft,*" even although "*the words in quotation
marks are taken from it verbatim.*"

These words, which I give in italics, (1) define
the purposes of the scheme to be "*for the social
and moral regeneration and improvement of persons
needy, destitute, degraded, or criminal, in some
manner indicated, implied, or suggested in the book
called 'In Darkest England.'*" Whence I apprehend that, if the whole funds collected are applied
to "mothering society" by the help of speculative
attorney "tribunes of the people," the purposes
of the trust will be unassailably fulfilled. (2)
The name is to be "*Darkest England Scheme,*" (3)
the General of the Salvation Army is to be
"*Director of the Scheme.*" Truly valuable information all this! But taking it for what it is worth,

the public must not be misled into supposing
that it has the least bearing upon the questions
to which neither I, nor anybody else, has yet been
able to obtain an intelligible answer, and that
is, where are the vast funds which have been
obtained, in one way or another, during the last
dozen years in the name of the Salvation Army?
Where is the presumably amended Trust Deed of
1888? I ask once more: Will Mr. Booth submit
to competent and impartial legal scrutiny the
arrangements by which he and his successors are
prevented from dealing with the funds of the
so-called "army chest" exactly as he or they may
please?

II. With respect to the "Eagle" case, I am
advised that Dr. Greenwood, whose good faith I
do not question, has been misled into misrepre-
senting it in the appendix to his pamphlet. And
certainly, the evidence of authoritative records
which I have had the opportunity of perusing,
appears to my non-legal mind to be utterly at
variance with the statement to which Dr. Greenwood stands committed. I may observe, further,
that the excuse alleged on behalf of Mr. Booth,
that he signed the affidavit set before him by his
solicitors without duly considering its contents, is
one which I should not like to have put forward
were the case my own. It may be, and often is,
necessary for a person to sign an affidavit without

being able fully to appreciate the technical
language in which it is couched. But his
solicitor will always instruct him as to the effect
of these terms. And, in this particular case,
where the whole matter turns on Mr. Booth's
personal intentions, it was his plainest duty to
inquire, very seriously, whether the legal phraseology employed would convey neither more nor
less than such intentions to those who would act
on the affidavit, before he put his name to it.

III. With respect to Mr. Bramwell Booth's
case, I refer the reader to p. 311.

IV. As to Mr. Booth-Clibborn's misrepresentations, see above, pp. 298, 299.

This much for the legal questions which have
been raised by various persons since the first
edition of the pamphlet was published.

DR. GREENWOOD'S "GENERAL BOOTH AND HIS CRITICS"

So far as I am concerned, there is little or
nothing in this *brochure* beyond a reproduction of
the vituperative stuff which has been going the
round of those newspapers which favour " General "
Booth for some weeks. Those who do not want
to see the real worth of it all will not read the

preceding pages; and those who do will need no
help from me.

I fear, however, that in justice to other people
I must put one of Dr. Greenwood's paragraphs in
the pillory. He says that I have " built up, on
the flimsy foundation of stories told by three or
four deserters from the Army " (p. 114), a sweeping
indictment against General Booth. This is the
sort of thing to which I am well accustomed at
the hands of anonymous newspaper writers. But
in view of the following easily verifiable statements, I do not think that an educated and, I
have no doubt, highly respectable gentleman like
Dr. Greenwood can, in cold blood, contemplate
that assertion with satisfaction.

The persons here alluded to as " three or four
deserters from the army " are :—

(1) Mr. Redstone, for whose character Dr.
Cunningham Geikie is guarantee, and whom it
has been left to Dr. Greenwood to attempt to
besmirch.

(2) Mr. Sumner, who is a gentleman quite as
worthy of respect as Dr. Greenwood, and whose
published evidence not one of the champions of
the Salvation Army has yet ventured to impugn.

(3) Mr. Hodges, similarly libelled by that un-
happy meddler Mr. Trotter, who was compelled
to the prompt confession of his error (see p. 277).

(4) Notwithstanding this evidence of Mr.
Trotter's claims to attention, Dr. Greenwood

quotes a statement of his as evidence that a
statement quoted by me from Mr. Sumner's
work is a "forgery." But Dr. Greenwood unfortunately forgets to mention that on the 27th
of December 1890 (Letter No. VII. above) Mr.
Trotter was publicly required to produce proof
of his assertion ; and that he has not thought fit
to produce that proof.

If I were disposed to use to Dr. Greenwood
language of the sort he so freely employs to me,
I think that he could not complain of a handsome
scolding. For what is the real state of the case ?
Simply this—that having come to the conclusion,
from the perusal of " In Darkest England," that
" General " Booth's colossal scheme (as apart from
the local action of Salvationists) was bad in
principle and must produce certain evil consequences, and having warned the public to that
effect, I quite unexpectedly found my hands full
of evidence that the exact evils predicted had, in
fact, already shown themselves on a great scale ;
and, carefully warning the public to criticise this
evidence, I produced a small part of it. When
Dr. Greenwood talks about my want of " regard to
the opinion of the nine thousand odd who still
remain among the faithful " (p. 114), he commits
an imprudence. He would obviously be surprised
to learn the extent of the support, encouragement,
and information which I have received from
active and sincere members of the Salvation

Army—but of which I can make no use, because
of the terroristic discipline and systematic es-
pionage which my correspondents tell me is enforced by its chief. Some of these days, when
nobody can be damaged by their use, a curious
light may be thrown upon the inner workings of
the organization which we are bidden to regard
as a happy family, by these documents.

THE SALVATION ARMY

ARTICLES OF WAR

To be signed by all who wish to be entered on the roll as soldiers

HAVING received with all my heart the Salvation offered to me by the tender mercy of Jehovah, I do here and now publicly acknowledge God to be my Father and King, Jesus Christ to be my Saviour, and the Holy Spirit to be my Guide, Comforter, and Strength ; and that I will, by His help, love, serve, worship, and obey this glorious God through all time and through all eternity.

BELIEVING solemnly that The Salvation Army has been raised up by God, and is sustained and directed by Him, I do here declare my full determination, by God's help, to be a true soldier of the Army till I die.

> I am thoroughly convinced of the truth of the Army's teaching.
>
> I believe that repentance towards God, faith in our Lord Jesus Christ, and conversion by the Holy Spirit, are necessary to Salvation, and that all men may be saved.
>
> I believe that we are saved by grace, through faith in our Lord Jesus Christ, and he that believeth hath the witness of it in himself. I have got it. Thank God !
>
> I believe that the Scriptures were given by inspiration of God, and that they teach that not only does continuance in the favour of God depend upon continued faith in, and obedience to, Christ, but that it is possible for those who have been truly converted to fall away and be eternally lost.

I believe that it is the privilege of all God's people to be "wholly sanctified," and that "their whole spirit and soul and body" may "be preserved blameless unto the coming of our Lord Jesus Christ." That is to say, I believe that after conversion there remain in the heart of the believer inclinations to evil, or roots of bitterness, which, unless overpowered by Divine grace, produce actual sin; but these evil tendencies can be entirely taken away by the Spirit of God, and the whole heart thus cleansed from anything contrary to the will of God, or entirely sanctified, will then produce the fruit of the Spirit only. And I believe that persons thus entirely sanctified may, by the power of God, be kept unblamable and unreprovable before Him.

I believe in the immortality of the soul; in the resurrection of the body; in the general judgment at the end of the world; in the eternal happiness of the righteous; and in the everlasting punishment of the wicked.

THEREFORE, I do here, and now, and for ever, renounce the world with all its sinful pleasures, companionships, treasures, and objects, and declare my full determination boldly to show myself a Soldier of Jesus Christ in all places and companies, no matter what I may have to suffer, do, or lose, by so doing.

I do here and now declare that I will abstain from the use of all intoxicating liquors, and also from the habitual use of opium, laudanum, morphia, and all other baneful drugs, except when in illness such drugs shall be ordered for me by a doctor.

I do here and now declare that I will abstain from the use of all low or profane language; from the taking of the name of God in vain; and from all impurity, or from taking part in any unclean conversation or the reading of any obscene book or paper at any time, in any company, or in any place.

I do here declare that I will not allow myself in any falsehood, deceit, misrepresentation, or dishonesty; neither will I practise any fraudulent conduct, either in my business, my home, or in any other relation in which I may stand to my fellow men, but that I will deal truthfully, fairly, honourably, and kindly with all those who may employ me or whom I may myself employ.

I do here declare that I will never treat any woman, child, or other person, whose life, comfort, or happiness may be placed within my power, in an oppressive, cruel, or cowardly manner, but that I will protect such from evil and danger so far as I can, and promote, to the utmost of my ability, their present welfare and eternal salvation.

I do here declare that I will spend all the time, strength, money, and influence I can in supporting and carrying on this War, and that I will endeavour to lead my family, friends, neighbours, and all others whom I can influence, to do the same, believing that the sure and only way to remedy all the evils in the world is by bringing men to submit themselves to the government of the Lord Jesus Christ.

I do here declare that I will always obey the lawful orders of my Officers, and that I will carry out to the utmost of my power all the Orders and Regulations of The Army ; and further, that I will be an example of faithfulness to its principles, advance to the utmost of my ability its operations, and never allow, where I can prevent it, any injury to its interests or hindrance to its success.

AND I do here and now call upon all present to witness that I enter into this undertaking and sign these Articles of War of my own free will, feeling that the love of Christ who died to save me requires from me this devotion of my life to His service for the Salvation of the whole world, and therefore wish now to be enrolled as a Soldier of the Salvation Army.

_____CORPS _____   18

....

........................*Corps.*

........................*Division.*

........................18

(SINGLE)

FORM OF APPLICATION

FOR AN APPOINTMENT AS AN

OFFICER IN THE SALVATION ARMY

Name ....

Address ....

1. What was your AGE last birthday ? .... What is
the date of your birthday ?.......................................

2. What is your height ? ....

3. Are you free from bodily defect or disease ?...................

4. What serious illnesses have you had, and when ? ....

5. Have you ever had fits of any kind ?..................... If so
how long ago, and what kind ? ....

6. Do you consider your health good, and that you are strong
enough for the work of an Officer ? .... If not,
or if you are doubtful, write a letter and explain the
matter. ....

7. Is your doctor's certificate a full and correct statement so
far as you know ? ....

8. Are you, or have you ever been married ?.....................

9. When and where CONVERTED ?...............................

10. What other Religious Societies have you belonged to ?.........

11. Were you ever a Junior Soldier ?... ... If so, how long ? ......

12. How long have you been enrolled as a SOLDIER ?............
and signed Articles of War ? ....

13. If you hold any office in your Corps, say what, and how
long held ?...

14. Do you intend to live and die in the ranks of the Salvation
Army ? ....

---

15. Have you ever been an open BACKSLIDER ?..............If
so, how long ? ....

16. Why ?...................... Date of your Restoration ?...........

17. Are you in DEBT ?...............If so, how much ?..............
Why ?...

18. How long owing ?...................What for ?......................

19. Did you ever use Intoxicating Drink ?...............If so, how
long is it since you entirely gave up its use ? ....

20. Did you ever use Tobacco or Snuff ?..................If so, how
long is it since you gave up using either ? ....

---

21. What UNIFORM do you wear ?...................................

22. How long have you worn it ? ....

23. Do you agree to dress in accordance with the direction of
Headquarters ? ....

24. Can you provide your own uniform and " List of Neces-
saries " before entering the Service ?........................

---

25. Are you in a Situation ?............If so, how long ?............

26. Nature of duties, and salary ....

27. Name and address of employer ? ....

28. If out, date of leaving last situation ? .... How
    long there ? ....
29. Why did you leave ?...
30. Name and address of last employer ?............................

---

31. Can you start the SINGING ? ....
32. Can you play any musical instrument ?......If so, what ?......
33. Is this form filled up by you ?...............Can you read well
    at first sight ? ....
34. Can you write SHORTHAND ? ....If so, what speed
    and system ? ....
35. Can you speak any language other than English ? ....If
    so, what ? ....
36. Have you had any experience and success in the JUNIOR
    SOLDIERS' WAR ? ....
37. If so, what ? ....

---

38. Are you willing to sell the " WAR CRY " on Sundays ? ...
39. Do you engage not to publish any books, songs, or music
    except for the benefit of the Salvation Army, and then
    only with the consent of Headquarters ? ....
40. Do you promise not to engage in any trade, profession, or
    other money-making occupation, except for the benefit of
    the Salvation Army, and then only with the consent of
    Headquarters ? ....
41. Would you be willing to go ABROAD if required ?............
42. Do you promise to do your utmost to help forward the
    Junior Soldiers' work if accepted ?............................
43. Do you pledge yourself to spend not less than nine hours
    every day in the active service of the Army, of which not
    less than three hours of each week-day shall be spent in
    VISITATION ?.................................. ....

44. Do you pledge yourself to fill up and send to Headquarters forms as to how your day is spent ? ....

---

45. Have you read, and do you believe, the DOCTRINES printed on the other side ? ....:...

46. Have you read the "Orders and Regulations for Field Officers" of the Army ? ....

If you have not got a copy of "Orders and Regulations," get one from Candidates' Department at once. The price to Candidates is 2s. 6d.

47. Do you pledge yourself to study and carry out and to endeavour to train others to carry out all Orders and Regulations of the Army ?...

48. Have you read the Order on page 3 of this Form as to PRESENTS and TESTIMONIALS, and do you engage to carry it out ?..

49. Do you pledge yourself never to receive any sum in the form of pay beyond the amount of allowances granted under the scale which follows ?..

ALLOWANCES—From the day of arrival at his station, each officer is entitled to draw the following allowances, provided the amount remains in hand after meeting all local expenses, namely :—For Single Men : Lieutenants, 16s. weekly, and Captains, 18s. ; for Single Women : Lieutenants, 12s. weekly, and Captains, 15s. weekly ; Married Men, 27s. per week, and 1s. per week for each child under 14 years of age ; in all cases without house-rent.

50. Do you perfectly understand that no salary or allowance is guaranteed to you, and that you will have no claim against the Salvation Army, or against any one connected therewith, on account of salary or allowances not received by you ? ....

---

51. Have you ever APPLIED BEFORE ?............ If so, when ? ....

52. With what result ? ....

53. If you have ever been in the service of the Salvation Army
in any position, say what? ....

54. Why did you leave? ....

55. Are you willing to come into TRAINING that we may see
whether you have the necessary goodness and ability for
an Officer in the Salvation Army, and should we conclude
that you have not the necessary qualifications, do you
pledge yourself to return home and work in your Corps
without creating any dissatisfaction? ....

56. Will you pay your own travelling expenses if we decide to
receive you in Training? ....

57. How much can you pay for your maintenance while in
Training? ....

58. Can you deposit £1 so that we can provide you with a suit
of Uniform when you are Commissioned? ....

59. What is the shortest NOTICE you require should we want
you? ....

60. Are your PARENTS willing that you should become an
Officer? ....

61. Does any one depend upon you for support? .... If so,
who? ....

62. To what extent? ....

63. Give your parents', or nearest living relatives', full address
....

_____

64. Are you COURTING? ....If so, give name and
address of the person ....

65. How long have you been engaged? ....What is the
person's age? ....

66. What is the date of Birthday? ....How long enrolled
as a SOLDIER? ....

67. What Uniform does the person wear? ....How long
worn? ....

68. What does the person do in the Corps ?...........................

69. Has the person applied for the work ? ....

70. If not, when does the person intend doing so ? ....

71. Do the parents agree to the person coming into Training ? ....

---

72. Do you understand that you may not be allowed to marry until three years after your appointment as an Officer, and do you engage to abide by this ? ....

73. If you are not courting, do you pledge yourself to abstain from anything of the kind during Training and for at least twelve months after your appointment as a Commissioned Field Officer ? ....

74. Do you pledge yourself not to carry on courtship with any one at the station to which you are at the time appointed ? ....

75. Do you pledge yourself never to commence, or allow to commence, or break off anything of the sort, without first informing your Divisional Officer, or Headquarters, of your intention to do so ?.....................................

76. Do you pledge yourself never to marry any one marriage with whom would take you out.of the Army altogether ? ....

77. Have you read, and do you agree to carry out, the following Regulations as to Courtship and Marriage ?.................

   (a) "Officers must inform their Divisional Officer or Headquarters of their desire to enter into or break off any engagement, and no Officer is permitted to enter into or break off an engagement without the consent of his or her D.O.

   (b) "Officers will not be allowed to carry on any courtship in the Town in which they are appointed ; nor until twelve months after the date of their Commission.

(c) " Headquarters cannot consent to the engagement of
    Male Lieutenants, until their Divisional Officer is
    prepared to recommend them for command of a
    Station as Captain.

(d) " Before Headquarters can consent to the marriage of
    any Officer, the Divisional Officer must be prepared
    to give him three stations as a married man.

(e) " No Officer accepted will be allowed to marry until
    he or she has been at least three years in the field,
    except in cases of long-standing engagements before
    application for the work.

(f) " No Male Officer will, under any circumstances, be
    allowed to marry before he is twenty-two years of
    age, unless required by Headquarters for special
    service.

(g) " Headquarters will not agree to the Marriage of any
    Male Officer (except under extraordinary circum-
    stances) until twelve months after consenting to his
    engagement.

(h) " Consent will not be given to the engagement of any
    male Officer unless the young woman is likely to
    make a suitable wife for an Officer, and (if not already
    an Officer) is prepared to come into Training at once.

(i) " Consent will be given to engagements between Female
    Officers and Soldiers, on condition that the latter
    are suitable for Officers, and are willing to come into
    Training if called upon.

(j) " Consent will never be given to any engagement or
    marriage which would take an Officer out of the
    Army.

(k) " Every Officer must sign before marriage the Articles
    of Marriage, contained in the Orders and Regulations
    for Field Officers."

PRESENTS AND TESTIMONIALS.

1. Officers are expected to refuse utterly, and to prevent, if possible, even the proposal of any present or testimonial to them.

2. Of course, an Officer who is receiving no salary, or only part salary, may accept food or other gifts, such as are needed to meet his wants ; but it is dishonourable for any one who is receiving their salary to accept gifts of food also.

THE DOCTRINES OF THE SALVATION ARMY.

The principal Doctrines taught in the Army are as follows :—

1. We believe that the Scriptures of the Old and New Testament were given by inspiration of God, and that they only constitute the Divine rule of Christian faith and practice.

2. We believe there is only one God, who is infinitely perfect, the Creator, Preserver, and Governor of all things.

3. We believe that there are three persons in the Godhead— the Father, the Son, and the Holy Ghost, undivided in essence, coequal in power and glory, and the only proper object of religious worship.

4. We believe that, in the person of Jesus Christ, the Divine and human natures are united, so that He is truly and properly God, and truly and properly man.

5. We believe that our first parents were created in a state of innocency, but by their disobedience they lost their purity and happiness ; and that, in consequence of their fall, all men have become sinners, totally depraved, and as such are justly exposed to the wrath of God.

6. We believe that the Lord Jesus Christ has, by His suffering and death, made an atonement for the whole world, so that whosoever will may be saved.

7. We believe that repentance towards God, faith in our Lord Jesus Christ, and regeneration by the Holy Spirit, are necessary to Salvation.

8. We believe that we are justified by grace, through faith in our Lord Jesus Christ, and that he that believeth hath the witness in himself.

9. We believe the Scriptures teach that not only does continuance in the favour of God depend upon continued faith in, and obedience to, Christ, but that it is possible for those who have been truly converted to fall away and be eternally lost.

10. We believe that it is the privilege of all believers to be "wholly sanctified," and that "the whole spirit and soul and body" may "be preserved blameless unto the coming of our Lord Jesus Christ." That is to say, we believe that after conversion there remain in the heart of the believer inclinations to evil, or roots of bitterness, which, unless overpowered by Divine grace, produce actual sin ; but that these evil tendencies can be entirely taken away by the Spirit of God, and the whole heart, thus cleansed from everything contrary to the will of God, or entirely sanctified, will then produce the fruit of the Spirit only. And we believe that persons thus entirely sanctified may, by the power of God, be kept unblamable and unreprovable before Him.

11. We believe in the immortality of the soul ; in the resurrection of the body ; in the general judgment at the end of the world ; in the eternal happiness of the righteous ; and in the everlasting punishment of the wicked.

DECLARATION.

I HEREBY DECLARE that I will never, on any consideration, do anything calculated to injure The Salvation Army, and especially, that I will never, without first having obtained the consent of The General, take any part in any religious services or in carrying on services held in opposition to the Army.

I PLEDGE MYSELF to make true records, daily, on the forms supplied to me, of what I do, and to confess, as far as I am concerned, and to report, as far as I may see in others, any neglect or variation from the orders or directions of The General.

I FULLY UNDERSTAND that he does not undertake to employ or to retain in the service of The Army any one who does not appear to him to be fitted for the work, or faithful and successful in it ; and I solemnly pledge myself quietly to leave any Army Station to which I may be sent, without making any attempt to disturb or annoy The Army in any way, should The General desire me to do so. And I hereby discharge The Army and The General from all liability, and pledge myself to make no claim on account of any situation, property, or interest I may give up in order to secure an engagement in The Army.

I understand that The General will not be responsible in any way for any loss I may suffer in consequence of being dismissed from Training ; as I am aware that the Cadets are received into Training for the very purpose of testing their suitability for the work of Salvation Army Officers.

I hereby declare that the foregoing answers appear to me to fully express the truth as to the questions put to me, and that I know of no other facts which would prevent my engagement by The General, if they were known to him.

Candidate to sign here ....

NOTICE TO CANDIDATES.

1. All Candidates are expected to fill up and sign this form themselves, if they can write at all.

2. You are expected to have obtained and read "Orders and Regulations for Field Officers" before you make this application.

3. Making this application does NOT imply that we can receive you as an officer, and you are, therefore, NOT to leave your home, or give notice to leave your situation, until you hear again from us.

4. If you are appointed as an Officer, or received into Training, and it is afterwards discovered that any of the questions in this form have not been truthfully answered, you will be instantly dismissed.

5. If you do not understand any question in this form, or if you do not agree to any of the requirements stated upon it, return it to Headquarters, and say so in a straightforward manner.

6. Make the question for this appointment a matter of earnest prayer, as it is the most important step you have taken since your conversion.

We must have your Photo.   Please enclose it with your forms, and address them " Candidate Department," 101, Queen Victoria Street, London, E.C

RICHARD CLAY AND SONS, LIMITED, LONDON AND BUNGAY.